Mattering Voices

Mattering Voices advances mutually enriching relationships between the transdisciplinary fields of voice studies and new materialisms.

This is the first edited volume to explore how the theoretical, methodological, and analytical possibilities of both voice scholarship and new materialisms gain further depth and directions through their co-constitutive—intra-active—relationality. In this book, voice researchers from performance studies and philosophy, artistic research, musicology, ethnomusicology, sound studies, feminist and gender research, and educational studies develop new materialisms-influenced approaches to voice and voice studies-inspired adaptations of new materialisms in the empirical study of various kinds of voicing. The topics covered range from voice in artistic practices and contemporary academia to new notions of musicality and vocal atmospheres, as well as the significance of singing in gendered senses of self and interspecies relations. By experimenting with intra-actions of voice studies and new materialisms, the book proposes fresh ways of researching and grasping how voices matter: how they materialize as events and practices and acquire meanings.

As a polyphony of voices, this volume invites readers into entangled conversations about how voice emerges—creating modes of being, knowing, and co-existing—and what it might still become.

Elisabeth Laasonen Belgrano is a Voice Artist, Performance Philosopher, Researcher (PhD) at the Academy of Art and Design University of Gothenburg, Existential Counselor, and Priest in the Church of Sweden.

Anne Tarvainen is an Ethnomusicologist (PhD) specializing in the embodied, experiential, and cultural meanings of voice. She is currently a Senior Research Fellow at the University of Eastern Finland.

Milla Tiainen is a Senior Lecturer in Musicology at the University of Turku and Associate Professor of Musicology (Title of Docent) at the University of Helsinki.

Routledge Voice Studies
Series editors: Konstantinos Thomaidis and Ben MacPherson

The *Routledge Voice Studies* series offers a platform for rigorous discussion of voice across disciplines, practices, and areas of interest. This series aims to facilitate the dissemination and cross-fertilisation of voice-related research to effectively generate new knowledge and fresh critical insights on voice, vocality, and voicing.

Beethoven and the Lyric Impulse
Essays on Beethoven Song
Amanda Glauert

Somatic Voices in Performance Research and Beyond
Christina Kapadocha

Owning Our Voices
Vocal Discovery in the Wolfsohn-Hart Tradition
Margaret Pikes and Patrick Campbell

Staging Voice
Michal Grover-Friedlander

Voice as Art
From Theatre to Forensics
Richard Couzins

Mattering Voices
Studying Voice through New Materialisms
Elisabeth Laasonen Belgrano, Anne Tarvainen, and Milla Tiainen

For more information about this series, please visit: https://www.routledge.com/Routledge-Voice-Studies/book-series/RVS

The Routledge Voice Studies series website can be found here: https://routledgetextbooks.com/textbooks/routledgevoicestudies/

Mattering Voices

Studying Voice through New Materialisms

Edited by Elisabeth Laasonen Belgrano, Anne Tarvainen, and Milla Tiainen

Routledge
Taylor & Francis Group
LONDON AND NEW YORK

First published 2025
by Routledge
4 Park Square, Milton Park, Abingdon, Oxon OX14 4RN

and by Routledge
605 Third Avenue, New York, NY 10158

Routledge is an imprint of the Taylor & Francis Group, an informa business

British Library Cataloguing-in-Publication Data
A catalogue record for this book is available from the British Library

ISBN: 978-1-032-48002-2 (hbk)
ISBN: 978-1-032-48829-5 (pbk)
ISBN: 978-1-003-39099-2 (ebk)

DOI: 10.4324/9781003390992

Typeset in Sabon
by KnowledgeWorks Global Ltd.

Support Material: //www.routledge.com/9781032480022

To Voice as Nothing, as All and as always More-than...

Contents

Contributors

Laura Cull Ó Maoilearca is a Writer, Scholar, and Artist based in the Netherlands working on the intersections of performance, philosophy, social justice, and human-nonhuman relations. Recent books include *Interspecies Performance* co-edited with Florence Fitzgerald-Allsopp. She is a Professor of Performance Philosophy at the University of Amsterdam and Lector for the Academy of Theatre and Dance, Amsterdam University of the Arts. She is a founding convener of the Performance Philosophy network and a co-editor of its journal and book series.

Jon Ivan Gill/Gilead 7 is an Assistant Professor of Philosophy at Gustavus Adolphus College, Minnesota, United States of America. Jon Ivan Gill uniquely engages the thought worlds of philosophy of religion, process philosophies, Hip-Hop philosophy, philosophy of race, poststructuralism, existentialism, and more through gravel-colored urban lenses. As Tomorrow Kings MC, Gilead7 ignites vinyl records and world stages with dystopian utopias of shattered spoken word.

Malou Juelskjær is an Associate Professor of Social Psychology at the Danish School of Education, Aarhus University, Denmark. Her work is situated within critical posthumanist, feminist new materialist thinking, currently centering on planetary, environmentally, and socially just knowledges and pedagogies. A co-founder of an organization that acquires agricultural plots to assist them in becoming future forests, wherefore she has added questions of multispecies care, soil ontologies, and re-wilding to her research focus – as well as experiments with what it may entail to listen to (the voices of) the more-than-human.

Elisabeth Laasonen Belgrano is a Voice Artist, Performance Philosopher, Researcher (PhD) at the Academy of Art and Design University of Gothenburg, Existential Counselor, and Priest in the Church of Sweden. Her research is dedicated to the art of vocal ornamentation, performative theology, and nothingness. She has been awarded a PhD in Performance (2011) for her practice-led voice research of nothingness through early modern operatic mad scenes and lamentations, and an MA in Systematic Theology (2024) with a thesis on 'Doing Theology *Through* Diffractive Methodology'. https://www.researchcatalogue.net/view/378762/430811

Tero Nauha is an Artist and Professor in Live Art and Performance Studies at Uniarts Helsinki, Finland. His research and artistic practice focus on the interstices between immaterial forms and material cultures, especially in the narratives that link financialization and its effects on the social sphere and knowledge production. He defended his doctoral thesis on schizoanalysis at the Theatre Academy of the Uniarts Helsinki in 2016. He was a postdoctoral fellow in the Academy of Finland-funded postdoctoral research project in 2016–2020 and a postdoctoral fellow at the Helsinki Collegium for Advanced Studies in 2017.

Norie Neumark is a Sound/Media Artist and Theorist based in Melbourne/Naarm, Australia. She is an Honorary Professorial Fellow at The Victorian College of the Arts, University of Melbourne, and Emeritus Professor, La Trobe University, Melbourne. She is the Founding Editor of *Unlikely: Journal for Creative Arts*. http://unlikely.net.au. Her collaborative art practice (www.out-of-sync.com) includes installation, new media works, and CD-Rom. Her current focus is human-animal relations, environmental concerns, and voice. Her publications include *Voicetracks: Attuning to Voice in Media and the Arts* (MIT Press, 2017) and *Voice: Vocal Aesthetics in Digital Arts and Media* (MIT Press, 2010), lead editor and contributor.

Jannie Pranger is a Mezzo-soprano who specialized in performing twentieth-century and twenty-first-century music. She began her singing career in 1986. Her farewell concert was in 2012. In 2020, she obtained her PhD at the Institute for Cultural Inquiry (ICON) of Utrecht University. She had a tenured teaching position at the School for Music and Technology at the HKU University of the Arts Utrecht. After retirement, she works as an independent lecturer and researcher. Her interests are the onto-epistemology, ethics, and aesthetics of music performance.

Julieanna Preston is an Artist, Writer, and Academic exploring the limits and potential in the vitality of materials, artifacts, landscapes, and built environments through durational site-specific live art, installations, videos, and performance writing. Based in Aotearoa/NZ, she presents and publishes internationally. www.julieannapreston.space

Mark D. Price is a multi-disciplinary Researcher with major interests in philosophy, arts, and music. He holds two PhDs, is the author of the two-volume biography of the artist and social activist Robert Lenkiewicz, several musical and dramatic works, and the ongoing 'No Self Can Tell' project, which has led to performances and presentations in Iceland, Sweden, Jerusalem, and Rome. He is currently based in England, at the University of Bolton.

Monika Rogowska-Stangret is an Assistant Professor at the Department of Philosophy and Cognitive Studies, University of Białystok, Poland. She is a philosopher conducting research at the intersection of feminist philosophy, environmental humanities, and critical posthumanism, and a translator. She is a recipient of a number of national and international grants and awards; most recently, the National Science Center in Poland supported her project *Anthropocene Ethics. Redefining the Concept of the Human in Posthuman Philosophy* (nr 2022/45/B/HS1/00849). She published, among others, in "Feminist Theory," "Philosophy Today," "The Minnesota Review: A Journal of Creative and Critical Writing." She is the editor-in-chief of "Matter: Journal of New Materialist Research."

Anne Tarvainen is an Ethnomusicologist (PhD) specializing in the embodied, experiential, and cultural meanings of voice. She is currently a senior research fellow at the University of Eastern Finland, working on the project "SOMECO—Sonic Mediations and Ecocritical Listening," funded by the Research Council of Finland. Her previous research has focused on embodied listening and singers' vocal expressions, somaesthetic experiences of d/Deaf, tone-deaf, and vocally disordered singers, as well as the transactions between voice, background music, and sonic environments in service industry workplaces. Tarvainen bridges academic insights with broader societal contexts, notably as a developer and educator of the Voicefulness® method.

Milla Tiainen is a Senior Lecturer in Musicology at the University of Turku and Associate Professor of Musicology (Title of Docent) at the University of Helsinki. She has published nearly 70 scientific articles on cultural studies of music, voice and sound, musical performance studies, and the methodological possibilities of new materialisms in the study of arts. Milla has co-edited special issues on these topic areas for journals such as *Body&Society* (2014), *Cultural Studies Review* (2015), and *The Polish Journal of Aesthetics* (2020), among others. She is a co-editor of *Musical Encounters with Deleuze and Guattari* (Bloomsbury Academic) and *New Materialism and Intersectionality* (Routledge).

Jennie Tiderman-Österberg is a Singer and an Ecomusicologist who is currently finalizing her PhD in musicology at The School of Music and Theatre, Örebro University, Sweden. Her research interests are posthumanism, new materialism, and multi- and trans-species musicking, and her doctoral thesis is a study of Nordic herding music as interspecies communication. She holds a position as an ethnomusicologist at *Dalarnas museum*, Sweden, and she is the artistic director and producer of the internationally renowned cosplay ensemble *Eldandili Fantasy Choir*. She is also the co-founder of the *International Network of Herding Music Scholars*.

Series Foreword

The claim that voice is everywhere might be a truism. Voice is predominant in interpersonal and technologically mediated communications and features prominently in discussions of identity, psychological development, and language acquisition. From theatrical performance to avant-garde or operatic singing, voice also offers aesthetic pleasure and, as is the case with rhetoric or journalism, it facilitates or imposes messages, arguments, and beliefs. Voice is also a powerful metaphor. Feminist scholars have championed the female voice, cultural studies have lent an attentive ear to subaltern voices, and the voice of the people is central to debates around politics, media, activism, and religion. In the arts, voice is not merely an instrument to be perfected or enjoyed. Notions of the artist's voice or, occasionally, the author's voice permeate relevant discourses. Non-human or posthuman voices invite us to listen to animal voices, interactive voice recognition systems, and vocal synthesis effected in robotics labs.

But how does one account for such plurality and multiplicity? How is voice to be discussed from a scholarly perspective? How might we move beyond bifurcated concepts of the voice in performance studies, for example?

The first, but decisive, step would be to create platforms for rigorous discussion of voice across disciplines, practices, and areas of interest. The *Routledge Voice Studies* series offers precisely such a platform. In the last few years, attention given to voice has shifted from sporadic publications in disparate areas of inquiry to the epicenter of discourses in a variety of overlapping disciplines. This series aspires to facilitate the dissemination and cross-fertilization of voice-related research and effectively generate new knowledge and fresh critical insights on voice, vocality, and voicing. To that end, we are delighted to include in the series of publications a variety of formats. We are equally interested in monographs, theme-edited collections, student-focused anthologies and sourcebooks, revised and expanded editions of classic texts, and inter-medial and multimedial outputs. Our hope is that these varied structures will attract both practitioners and scholars as contributors and find a readership among established and emergent researchers, students, and artists.

We understand voice studies as a shifting landscape of questions and concerns, as a proliferative inter-discipline. Building on current initiatives, we wish to expand and capitalize on the productive debates taking place in the areas of music, theater, and performance studies, as well as cultural studies, ethnomusicology, sound studies, acoustics, and acoustemology. Yet we are equally as keen on extending an invitation to inputs from psychology, fine art, poetics and orality studies, linguistics, media and film studies, robotics and artificial intelligence, history and philosophy, translation and adaptation studies, among others. Spearheaded by the discussions across disciplines and cultures hosted in its inaugural publication, the edited collection *Voice Studies: Critical Approaches to Process, Performance and Experience*, this book series listens out for new spaces in which voice can reverberate with revitalized vigor. We hope you enjoy this fascinating journey with us.

Series editors: Dr Konstantinos Thomaidis
and Dr Ben Macpherson

Foreword

"Mattering Voices"

Jon Ivan Gill

Years ago, while doing an underground rap project in a secluded studio in the suburbs of Chicago, I was given three technically superb instrumentals by a beatmaker. Every kick, snare, stolen sample (yes, we do this in the subverse and rarely ask for permission), and arrangement was quite pristine and inviting as a space on which to place the nihilistic infinitude of my neo-Nietzschean voice. Nothing was offbeat, sloppily placed, or haphazardly done. It was quite clean. Maybe this was the problem. Perhaps it was too clean. I put all three of these bangin' instrumentals to use and produced an album fertilized with rhythmic neo-beatnik (shout to Chicago veteran MC Offwhyte) phrasings that I don't even have physical copies of now and occasionally sell in Spain and eBay every now and then for no less than $80. Still, there was something missing. The swing of the natural world that embodies the production of Hip-Hop instrumental legends such as Pete Rock, Madlib, and the late J Dilla, the swing that makes present the past of Duke Ellington, Perez Prado, and Willie Hutch, was absent. The instrumental itself was a series of time signatures that could easily be deconstructed by any musicologist or enthusiast of sound and lauded for its technical mastery. But art enshrines and venerates the soul. This is not the essential entity that Descartes presupposed transcended the material this volume is so centered on. Rather, it is the ever-fleeting moving pieces of cultures, histories, pains, pleasures, disabilities, and new world architectures in process which flow through all of us in that thing we stutter to call "subjectivity." This was absent from the instrumental. Walter Benjamin's aura was lost and replaced with the mechanical reproduction of ancient sounds, as bland as the postcard of the mountain robbed of its authenticity when it was distributed all around the world. In either the attempted transformation of practice to theory or from the theory to an assumption of practice, we risk the loss of the dynamism of that which oozes outside of the formula. In other words, a careless oscillation between theory and practice, serving as an arm of colonial and post-colonial coercion from the brutal theft of land to the commodification of the human into racial hierarchies for capitalist empires, can be systematized not only to exclude but also preclude that which *matters* in aesthetic expression. This novel and audacious volume of open-ended scholarship, *Mattering Voices*, empowered by *poiesis*, plays at the InnerSection of voice studies and new materialisms and offers several means to rethink voice by reimagining what it is, does, and is relevant to.

When I say "InnerSection," I refer to an alternative approach to the now classic and established notion of "intersection," or a critical attunement to how the linguistic systems within which we identify and are shaped entangle one positionality with several others that can appear in one human. Because of intersectionality, the oppression of someone who may be referred to in some social contexts as a "White" "Woman" might be remarkably different than that of someone who identifies and is identified as a "Woman" and also as "Black" within a systemic structure that is both racist and sexist. In some ways, InnerSection benefits from the acknowledgment of the entanglement in theories of intersectionality and an analysis of power. But the aforementioned beat was the product of a well-thought-out analysis of how one should construct an instrument suited for the facet of Hip-Hop Culture some refer to as "independent rap." Yet, it, like intersectionality, fails to experience the precursor to the arbitrary line drawn that is, in many ways, a preliminary point of origin of the existence of the positionalities of gender and race. In other words, the "intersection" arguably presupposes that race and gender are "real" distinctions (read necessarily existing whether or not they appeared unmediated by the linguistic categorization found practiced by those classified as "homo sapiens"). If we think in terms of independently existing "things," then the means of coming into relation must be through an "intersection," a type of collision of two or more "things." However, InnerSection invites us to rethink if the ways in which we have determined "this" from "that" are the most helpful, especially since it is possible to think of the universe (perhaps more fruitfully) aside from the arbitrary (and in many instances intrinsically oppressive) concepts of race, gender, nationality, disability, insanity, and so forth. What if that which we have engaged through the lenses of intersection even in the interests of more equitable futures is more adventurously thought of if voice, privilege, oppression, history, discovery, war, peace, and so much more are all conceived not as static unchangeable and unrelated *things* but as open-ended *processes* in an ever-evolving multiplicity/manifold? In many instances, we assume that there are separate and virtually unrelated entities (the intersectional identity of an "Indigenous" "Woman" over here against the intersectional identity "Latina" "Woman" over there). However, what presents and futures would emerge if we were to enact a collapse of significant arbitrary distinctions such as race and gender that invites an engagement of the loss of difference in relation (the InnerSectional misplacement of categories "Latina," "Indigenous," and "Woman" and their connotation in any given society)? Would this give way to a perpetual theopoetic humanism co-creating the next steps with the fluidity that was trapped behind the ossification of the aforementioned stagnant and ostensibly hierarchical categories?

Mattering Voices is an example of what InnerSection is and can be amidst the transdisciplinary studies of new materialisms and voice studies. Elisabeth Laasonen Belgrano, Anne Tarvainen, and Milla Tiainen have curated a set of daring essays that challenge us, the readers, to be persuaded by what appears when the phenomenology of the voice strikes us in the instant, reworking the application of theory to embody and not reduce novel manifestations that do not fit but diffract around our current articulations. In the spirit of the InnerSection, the editors and writers are careful to emphasize theory and the material world as not opposing, different, and unrelated realms, only capable of connecting at an "intersection" in the space of the detached Cartesian observer. Rather, all of the essays strive to think of theoretical approaches and the material Sartrean "nausea" upon which they think through as two conversation partners made of cigar smoke, informing and enfolding within each other as the distinguishing of one from the other becomes unthinkable ... and undesirable. In these free explorations of this volume, we are lured into thinking of voice as the human and non-human material of the world we always intuited it could be. We are privileged—dare I say "blessed"—to have the text's poetic energy from which to create our own meanings on various iterations of voice, the energy which we emote, hear, and feel and cannot locate spatially but are inevitably transformed through perpetually.

The three beats were ossified, yes. But I injected live words created by the material residue of the South Side of Chicago, Belize, the Blues, Calypso, El Trains, and football (soccer). In the words, I said the poetic difference that Hegel hinted at, and Heidegger could not utter. *Mattering Voices* matters because in it we are moved to magnify the miniscule into massive modules for progressive movement to move our musings on material from massively inert and moored to the multiplicitous manifold of meaning-making.

Introduction

How to let a crossroads emerge between (the study of) voices and new materialisms?

Milla Tiainen, Elisabeth Laasonen Belgrano, and Anne Tarvainen

> Placed in a multi-dynamic landscape surrounding us on all sides, we close our eyes and allow ourselves to let go of thoughts. To let go of that which we know. Of what we feel. The task may seem impossible because of what we have been trained "to do" since we were young. But let us just begin to imagine ourselves. As breathing bodies in relation to air. Perhaps we may hear the sound of other voices. Speaking. Singing. Connecting. From afar. From near. From balconies. From fear. From relief. From any climate. From in-between living and dying ....[1]

Among the many new or reconfiguring sociocultural phenomena that characterized the exceptional and often-tragic times of the COVID-19 pandemic, some new forms of voicing also came into being. One such form, which received international attention and became known as "balcony singing," first emerged across Italian cities and towns from Naples and Rome to Florence and Turin. In the early stages of the pandemic in 2020, parts of Italy were particularly hard hit by the coronavirus by European and even global standards, with a high number of infections and fatal cases. In an attempt to get the first wave of the pandemic under control through sociopolitical regulation, Italy's inhabitants were subjected to strict weeks-long lockdown measures (Tondo 2020). Devising an idiosyncratic way to cope with these unprecedented restrictions, the quarantined residents of apartment buildings and tower blocks started to sing from their balconies and windows. Many people also played instruments or DJ sets. Ranging repertoire-wise from Italian folk songs and opera arias to renowned popular music tunes such as "Volare," the proliferating acts of lockdown singing unfolded as a way to express solidarity and animate a sense of community among humans under deeply difficult circumstances. Soon, the trend of balcony singing spread from Italy to other European countries. This was aided by invitations to sing and play and video footage on the earlier performances that circulated in social media and via journalistic online articles.

There would be many ways to delve deeper into the phenomenon of balcony singing in research concerned with voice; or with multifarious, historically, socioculturally and materially contingent practices of voicing. Leaving strictly physiological and acoustical considerations of vocal sound aside, one could approach balcony singing by drawing on a wealth of psychological,

DOI: 10.4324/9781003390992-1

music and health, and educational studies into the beneficial effects of group singing on the participants' biopsychosocial well-being (Norton 2016; Irons and Hancox 2021: 11–12). From a notably different perspective, one could explore whether the media coverage of these performances originating in Italy perhaps reiterated the long-standing cultural stereotype of Italians as lovers and skillful producers of beautiful singing voices and vocal music (Dolar 2006: 3–4). Importantly, one could also elaborate on numerous concepts and research approaches that have been developed in the vibrant field of inter or transdisciplinary[2] voice studies during the past 10–15 years. Possible approaches would range from close analysis of the (recorded) balcony performances – their vocal and bodily, multisensory features – to theorizing them as an *intermaterial* practice occurring between embodied vocalizers and audiences and other media of sound, such as air (see, e.g., Eidsheim and Meizel 2019: xxv; Eidsheim 2015: 1–25).

All the approaches just envisioned, and no doubt many others, would provide fruitful insights into the pandemic-era voicings in question. Simultaneously, each approach would result in a necessarily limited perspective on manifestations of voice within balcony singing. Accepting the partial nature of any research angle and methodology, we propose there are yet further aspects to be discovered in the balcony singing performances. Attention to these aspects may enrich voice research especially in terms of acknowledging the *manifold, active and relationally forming materiality* of voicings, which the balcony vocalizations throw into sharp relief. Here, we are inspired by physicist and feminist philosopher Karen Barad's proposition (2007: 139) that phenomena consist of "*the ontological … entanglement of intra-acting agencies.*" The term intra-action suggests that the different agencies making up a phenomenon do not merely *inter*act with each other as originally ontologically separate entities. Arguably, the term intermaterial mentioned above gestures to a well-established concept of interaction. Instead, intra-action posits that the very existence of entities is shaped by their mutually co-constitutive relations. Approached from this perspective, the phenomenon of lockdown balcony singing, too, is composed of various intra-acting agencies, many of which are material.

To begin with, one cannot ignore the agency of the coronavirus or its particular early variant within the balcony performances. Because of its serious but at first murkily understood threat to human life, the coronavirus shook the illusion – still often well and alive – of humans as a species distinct from and in control of other life forms and nature. The virus highlighted the openness and vulnerability of human bodies to diverse material agents – such as this microbe, which travels via air between humans and possibly other species, while also spreading with the aid of airplanes, tube ways and other forms of globalized transport technology. Quite soon after the pandemic had begun, it was understood that human acts of voicing could enhance the spreading of the coronavirus. Consequently, the virus's potential transmission via droplets and aerosols when people cough, talk or otherwise vocalize in each other's vicinity suddenly rendered scenarios such as shared situations of singing a health hazard. However, the coronavirus's material capacity for infecting human organisms also fed people's yearnings for new, danger-free forms of being together. In part, these yearnings concerned the possibility to keep enjoying the shared time-spaces and sense of connection afforded by vocal sounds and music. The eventually transnational events of balcony singing unfolded as one actualization of such strivings. At the same time, these performances demonstrated the power of voicings as physically moving and touching vibrations, sensorial and symbolic acts, and processes that may transport their participants toward novel relations and experiences.

The materialities intra-actively contributing to balcony singing also included the involved urban structures. Besides telling things about urban planning policies, architectural styles, and the residents' standard of living, the buildings from which the lockdown voicings sounded co-constituted these performances in sheer material terms. The buildings' size, construction materials and spatial location all shaped the environmental movement and acoustics of the vocal emissions and their attendant capacity to affect the vocalizers and their audiences (see Goodman 2010: xviii–xix, 75–79). Lastly, the material agency of media technologies must be considered as significant here. This applies, above all, to the recording, distribution and modulative capacities of digital technologies. The contributions of these technologies ranged from video clips of the performances made and shared with the residents' smartphones to fake versions of these clips, in which the original performances were overdubbed to make it seem the performers were singing Anglo-American pop music hits (Rowley 2020).

In sum, the multiple actively participating materialities of balcony singing, extending from human and non-human bodies, bodily produced sounds and organic fluids to air currents and portable networked devices, intra-acted with each other to jointly give shape to these voicings. Such material processes subsequently intra-acted with the ways in which different recipients could grasp the performances at the level of symbolic meanings and social discourses.

Inviting intra-actions of voice studies and new materialisms

This edited volume, *Mattering Voices*, springs from the premise that in order to enrich understandings of how voices matter – how they materialize into diverse events and practices and come to mean different things – it is fruitful to enhance the interplay, or to try and generate an intra-active crossroads, between transdisciplinary voice studies and *new materialisms*. While our opening example above concerned singing, as a whole, this volume addresses an array of voicing processes: speaking, calling, narrating, music-making, stuttering, improvisatory, interrupted, yet-to-emerge, contemporary, historical, technologically informed, machine-generated, human as well as more-than-human.[3]

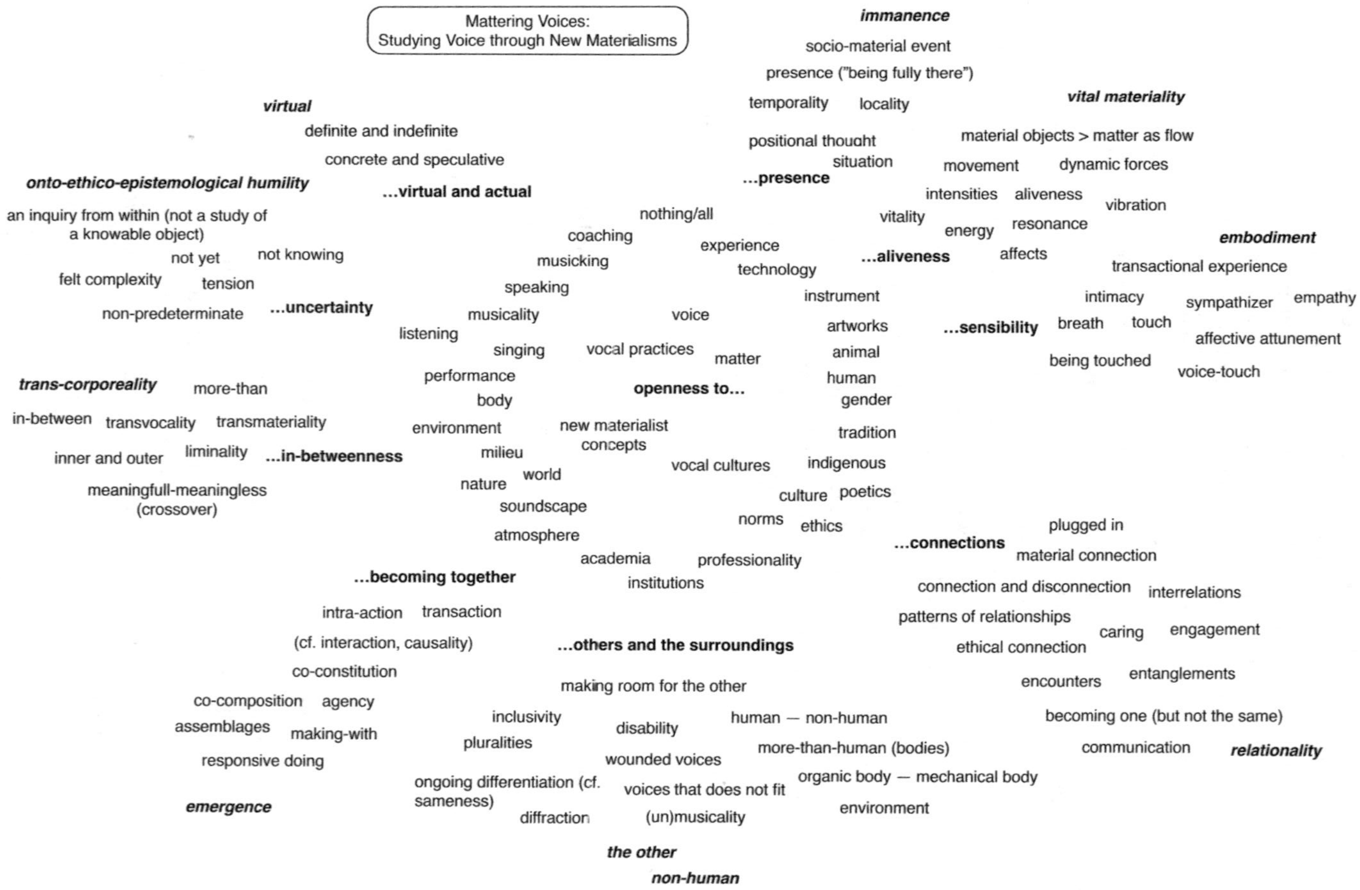

Figure 0.1 Mind map for *Mattering Voices*.

Source: Developed by Anne Tarvainen in collaboration with Milla Tiainen and Elisabeth Laasonen Belgrano.

By transdisciplinary voice studies, we refer to ways of studying the myriad relations, performances, experiences, knowledges, concepts, and roles of voices or voicings that have burgeoned across artistic research, the humanities and the social sciences, especially since the 2010s. One important driving force behind the flourishing of voice research has been the Routledge Voice Studies book series in which the present collection is included. As will be discussed further in the next sections, by new materialisms, we refer to an equally transdisciplinary assemblage of theories and research approaches across the arts, humanities and social sciences. To underline the range of approaches that belong to new materialism, many researchers have chosen to use the plural form "new materialisms" (Coole and Frost 2010a). This form is used in *Mattering Voices* for the same reason. Yet, what unites these approaches is their reconsideration of the nature and significance of materialities as a constitutive part of human realities and the world beyond the human. Like voice studies, new materialisms have proliferated especially since the 2010s, although their beginnings (similarly to voice research) can be traced back to scholarship carried out in the 1980s–2000s and also much earlier.

The intra-actions of voice studies and new materialisms advanced in this volume rest on a twofold idea. On the one hand, we maintain that both research areas, although lacking fixed borders, display substantial distinctive features. Both are characterized by a range of specific research themes and interests, theoretical influences and methodological considerations. Both also involve accumulated in-depth knowledge on their respective main topics: the phenomenon and concept of voice in the case of voice studies and materiality as something that acts and makes a difference – matters – in the case of new materialisms. On the other hand, voice studies and new materialisms are understood here as permanently emergent fields, or as shifting landscapes of questions and concerns, to paraphrase how the editors of Routledge's Voice Studies book series, Ben Macpherson and Konstantinos Thomaidis, describe voice research on the series' webpage. In particular, we propose that when elements from new materialisms and voice studies enter into new *relations* – be it with novel subject matters or theoretical possibilities – this enables these elements and scholars working with them to re-begin: to formulate questions about voices and materialities from a slightly or considerably transformed perspective. Such new relations may also renew the meanings and uses of the concepts, theories, and methodological approaches already established within voice studies and new materialisms, respectively.

Committed to such re-beginnings, all the chapters of *Mattering Voices* explore how the theoretical, methodological, and analytical possibilities of both voice studies and new materialisms can be expanded through their intra-active relationality. In this collection, researchers from several fields, which span, alongside voice studies, artistic research, musicology, ethnomusicology, performance studies and philosophy, sound studies, feminist and gender research, and educational studies, develop new materialisms-inflected approaches to voicings and voice studies-influenced adaptations of new materialisms. The volume's chapters probe the usefulness of new materialist perspectives for constantly asking what voices may be, what they can do, and how each research approach allows for particular ways of comprehending and studying voice while impeding others (cf. Eidsheim and Meizel 2019: xiv–xv). At the same time, the chapters demonstrate how the rich insights into the voice's cultural, sociopolitical, ethical and material dimensions developed within voice studies help to test and expand the ideas of materiality and agency and the process and relation-based ontological stances that are integral to new materialisms. In this manner, the contributions to *Mattering Voices* make a plurivocal claim for the mutual relevance of – always-emergent – voice studies and new materialisms. All the chapters anchor the intra-action of these fields in empirical study of specific voicing activities, cultures, and experiences.

As such, considerations of materiality are nothing new to voice research. The relationship of voicings to the human bodies of vocalizers and other participants of vocal events has grown into a rich theme since such early discussions as Roland Barthes's (1977) vastly cited essay on the bodily grain of the voice. During the past several decades, explorations of voice-(human) body relations have focused on the entanglement of corporeal materiality with the symbolic messages of speech, song or other voicings, the culturally contingent (gendered, racialized, class-bound, and so on) bodily features and capacities of voicers and voice perceivers, and the changing spatial and technological dimensions of voice-body coexistence (Dunn and Jones 1994; Connor 2001; Cavarero 2005; Neumark et al. 2010; Bosma 2013; Macpherson and Thomaidis 2015; Eidsheim 2019; Rosa 2019; Couzins 2023; see also the volumes of *Journal of Interdisciplinary Voice Studies*). Today, inquiries into the material aspects of voice cover, then, a vast array of perspectives. These extend from studies of voice as somatic modes of sounding, listening and writing (Tarvainen 2018; Kapadocha 2021a) to posthumanist approaches to how voices operate in a world always co-created by material beings more-than-human, ranging from technologies to non-human species (Belgrano 2015; Neumark 2017; Tiainen 2018; Bentivegna 2019; Thomaidis 2019a; Butler 2021; Bentivegna and Edlund 2022).

A couple of books have already mapped the specific crossroads of voice research and new materialisms. Norie Neumark's *Voicetracks* (2017) is, to our knowledge, the first monograph that examines – in the context of recent decades' voice and media art – what writing about voice can offer to new materialist thinking and vice versa. Virginie Magnat's *The Performative Power of Vocality* (2019) includes a wonderfully rich chapter titled "Exploring (K)new Paradigms" that discusses new materialist theories, related fields from affect studies to neuroscientific research, and how to engage these largely Western approaches in a respectful, decolonizing dialogue with Indigenous ontologies, epistemologies and methodologies relevant to voice research. Moreover, there are many articles and chapters, published since the late 2000s, in which scholars across the humanities and social sciences elaborate new materialist theories in the study of diverse voicing practices (Tiainen 2007, 2013, 2017;

Mazzei 2016; Mazzei and Jackson 2017; Leppänen 2018; Leppänen and Tiainen 2018; see also Tiainen 2012; Laasonen Belgrano 2018, 2019, 2021, 2025; Uehara and Laasonen Belgrano 2020).

The ways in which *Mattering Voices* expands this body of research are linked to the status of the volume as the first essay collection on intra-actions of voice studies and new materialisms. Because this is an edited book, it accommodates researchers from various scholarly and cultural backgrounds. All have, however, worked for years through new materialist perspectives in their research on voice. When these contributors delve into the same or closely connected concepts – for example, agential matter, intra-action, trans-corporeality and diffraction – while navigating different subject matters and research traditions, this showcases the diversity of manners in which such new materialist concepts can help make sense of vocal phenomena. Indeed, we hope that this assembling of many projects into one volume will create synergy between their ways of putting voice scholarship and new materialisms in intra-action. Such synergy may emerge when readers explore the chapters in varying orders, making connections across them according to their distinctive interests and reading tactics. Besides aspiring to this kind of synergy, *Mattering Voices* includes research on many topics previously unexplored at the crossroads of voice studies and new materialisms. Among others, these topics include amateur and professional singers' experiences of acid reflux, investigated via an interplay of new materialist and pragmatist philosophical perspectives (Chapter 4) and gendered voices in contemporary neoliberal academia (Chapter 5).

To sum up the main aim and the ethos of this volume, we hope to cultivate intra-actions of voice studies and new materialisms as *open-ended means of doing research*. We agree with the critiques according to which any assumption that new materialisms would represent something wholly novel – as potentially implied by the term "new" – is untenable. Such assumptions fail to acknowledge the multifaceted considerations of materiality in earlier feminist, social and art theories (Ahmed 2008) – and indeed in voice studies. Thus, the present book does not propose that new materialisms should come and replace other approaches to materiality and associated concerns in voice studies (or elsewhere). Instead, new materialisms provide *some* fresh concepts and potential for understanding the materialities of voicings, the nature of vocal agency and relations, and the interconnectedness of ontology, epistemology, and ethics in vocal activities. How this potential of new materialisms is actualized depends on the connections and aspirations of each study. Therefore, *Mattering Voices* addresses new materialisms – as well as any concern and concept of voice studies – as reiterative but always with a difference and as open to reconfiguration in changing relations. Such relations characterize this collection starting from the mutually different backgrounds of the editors, which span combinations of artistic research, musicology, ethnomusicology, voice studies, gender studies, (Western) classical and operatic singing, singing-songwriting, popular music, performance philosophy, theology, spiritual and liturgical voicing, voice coaching, and improvisation.

Encouraging surprisingly sprouting connections, we propose that the word "through" in the volume's subtitle, "studying voice through new materialisms," signals the and ... and ... and (Deleuze and Guattari 1987: 119) or the more-than of new materialisms which is enabled by heterogeneous ways of engaging with new materialist approaches. When differently oriented voice researchers and readers think, feel and practice *through* new materialisms, these processes keep expanding the ways in which such modes of exploration materialize and the other elements with which they do this. Thereby, new materialist concepts and orientations continue to (co-)generate ways of researching voice and tangible ways of sensing, producing and experiencing voices.

Proposals for defining new materialisms

To further illuminate the focus and aspirations of *Mattering Voices*, it is important to clarify how the term "new materialisms" is understood here. A useful starting definition is that new materialisms refer to multiple cross-disciplinary theories and research approaches that study materialities as a dynamic and efficacious, or *agential*, part of human practices and the world beyond the human. The materialities that matter to new materialist research are distinctly plural. They span the activities and substance of human and other animal bodies, the entities and forces of the natural world, from microorganisms to climates, as well as human-made artifacts and the material, socioeconomic conditions of living. The most known and cited sources associated with new materialisms tend to be certain works in philosophy, feminist and social theory and science studies written largely by Anglo-American researchers (DeLanda 2002; Barad 2007; Alaimo and Hekman 2008; Alaimo 2008; Alaimo 2010; Coole and Frost 2010a; Bennett 2010; see also, e.g., Braidotti 2002, 2013). Many contributors to *Mattering Voices* also engage with these works. Yet, we want to emphasize the varied disciplinary and geographical roots of new materialisms. Since the 1990s and 2000s, their beginnings can be found, at least, in feminist theory, political theory, and science and technology studies in Europe and the US (Haraway 1991, 2016; Hird 2004; Kirby 2011); media theory inspired by research in the German-speaking academy (Kittler 1999; Parikka 2010; Ernst 2012); and studies of art developed in Australia, the Nordic countries, and other parts of the world (Tiainen 2007; Kontturi 2009; Kontturi 2018; Barrett and Bolt 2013).

The various roots of new materialisms manifest themselves in this collection. The contributors do voice research led through new materialist perspectives in several of the fields above, for example, the study of arts or feminist and gender research, while hailing from diverse locations, including Central and Northern Europe and Australia. Despite this diversity, new materialisms betray a Euro or Western-centric undercurrent, which we will critically consider below.

A key common feature of new materialisms that makes the label feasible is the premise that it is crucial for scholars in the humanities and social sciences to re-examine how different materialities, as outlined above, actively co-constitute cultures and societies, or human and more-than-human modes of being. In promoting the idea that materialities have agency, new materialist theories challenge a long-standing tradition of marginalizing and devaluing the material world in mainstream Western philosophy since antiquity in favor of other categories, such as mind and human intentional subjectivity (understood as separate from matter). At least two reasons have been offered for what makes a revived attention to materialities necessary both historically and at present. First, several developments of recent decades, such as the dynamic understandings of materiality in late 20th-century and 21st-century natural sciences, the entanglement of human lives with digital and biotechnologies and the global material processes of the climate crisis, require that humanist and social scientific scholars also reconsider the nature and impact of matter and the related onto-ethico-epistemological, political concerns (Coole and Frost 2010b: 5–36; Braidotti 2013; Gamble et al. 2019). Second, especially in the formative stages of the concept of new materialism, these approaches were presented as an alternative to the arguably overstated role of language, discourse and symbolic representation in theories concerning the socially constructed character of reality (Alaimo and Hekman 2008: 1–4; Van der Tuin and Dolphijn 2010). Famously, social constructionist approaches had shaped the humanities and the social sciences since the 1960s, achieving by the 1980s–1990s a dominant status in many fields. Both arguments just outlined pertain to what has been seen as the novelty of "new" materialisms (Coole and Frost 2010b: 4–15; Dolphijn and van der Tuin 2012: 25).

Mattering Voices endorses the view that new materialist approaches have complemented and reconfigured social constructionist stances, for instance, by examining the intra-active relations of discourse and matter within scientific, artistic and other practices (Barad 2007; Arlander 2014; Golovátina-Mora et al. 2022). However, as indicated earlier, this collection does not regard new materialisms as an absolutely novel step in research or as a corrector of previous flawed perspectives. This is reflected in how the chapters included herein do not proceed by contrasting new materialisms with social constructionist or some other theories. Rather, they explore what modest, yet significant, fresh understandings of voicings can be generated *with* new materialist ideas, *in relation to* other theories and research approaches. Thus, in this volume the newness of new materialisms means, above all, the shifting explorative powers of these perspectives that unfold along with each study in which new materialisms are being elaborated in particular, situated ways (cf. Tiainen et al. 2015: 5; Laasonen Belgrano and Price 2025a). The chapters of *Mattering Voices* present a variety of such ways, ranging from a study of voice as a symbol of (intra-acting) Nothingness in 17th-century Venetian opera, following traces of vocal praxis leading to the historical fields of theology and alchemical sciences, and how the notion of nothingness can be revived in contemporary artistic voice research (Chapter 1), to a study of how Western classical singing and other genres of vocal music afford experiences of self for a transmasculine singer (Chapter 3).

Besides their cross-disciplinary roots and changing adaptations, new materialist approaches also vary in that some of them put more emphasis on the agential capacities inherent to matter (DeLanda 2002; Bennett 2010). Other approaches are known for insisting that materialities act and acquire their characteristics always in an intra-active relation with other material, symbolic, and social processes (Bolt 2004; Barad 2007). In the case of most new materialist outlooks, this difference of emphasis is a matter of degree instead of a fundamental divergence, and both stances on the agency of matter may be emphasized in one and the same study. The simultaneity of or subtle variation between these emphases can be seen in the chapters of *Mattering Voices*, too (e.g., Chapter 7).

One theoretical strand that stresses the irreducible realness of the material world, but which is almost entirely absent from this collection, is object-oriented ontology (OOO). According to several advocates of new materialist thinking (Åsberg et al. 2015; Tiainen 2018), the perspectives of OOO differ from new materialisms because of their more traditionally realist premise that material reality leads an independent existence, ultimately residing beyond the reach of human (and non-human) cognition. Another characteristic argument of OOO approaches is that the relations of objects to other entities are secondary to these objects' innate essence (Bryant 2011; Harman 2011). There is a difference, then, between understanding material entities as existing in a relational manner within most new materialist theories and the emphasis of OOO on the material world's self-contained core. The contributions to *Mattering Voices* build strongly on the process and relation-driven ontological stances of new materialisms. This explains the scarcity of engagement with OOO.

Yet, another aspect of the plurality of new materialisms discernible in the present volume is that due to the cross-disciplinary emergence of these perspectives, their preoccupation with materialities is motivated somewhat differently depending on where new materialist theories are being developed. For example, in feminist and gender research, one of the pioneering areas for new materialisms, reconsiderations of materiality have been propelled by the perceived need to rethink how the material features of bodies and nature truly matter in gendered, racialized, class-related, and other forms of lived social experience. New materialist dynamic understandings of matter enable this rethinking without invoking essentialist ideas about the material world as the fixed origin of gender or other identity markers. Attention to the agential materialities of human and non-human existence also challenges the Western human-centered and colonialist worldviews. These views have been deeply harmful to natural environments and human cultures, and feminist thinkers have demonstrated that they are premised on the prioritization of white men (Haraway 1991, 2007; Braidotti 2002, 2013; Åsberg and Braidotti 2018).

In social and political research, a revived interest in materialities has been motivated by the conviction that adequate accounts of human societies and political processes require an expanded focus that acknowledges the role and impact of non-human material actors in social realities (Latour 2007; Tsing 2015; Alaimo 2016). This conviction continues to grow stronger in the current age of anthropogenic ecological emergency. Finally, new explorations of matter in the study of arts have been driven by the perception that the poststructuralist understandings of art prevalent since the 1980s, in terms of "texts" through which cultural representations and discourses are (re)produced, overshadow the more material yet equally significant aspects of artistic production and experience. There is no return to the mystified Romantic notions of artistic creativity and autonomous artwork in these new materialist perspectives. Instead, they examine art – whether visual, musical, sound, or performance art – as a wealth of simultaneously materially and socioculturally bound activities (Tiainen 2007; Kontturi 2009; Barrett and Bolt 2013; Hickey-Moody and Page 2015; Belgrano 2016; Gatt 2017; Ingold 2022; Kontturi and Tiainen 2024).

The disciplinary diversity of new materialisms and their motivations permeates *Mattering Voices* while being further enriched by voice studies perspectives. Several chapters engage with feminist new materialist concepts, such as trans-corporeality, first formulated by the environmental humanities scholar Stacy Alaimo. This concept and a related notion of touch are expanded upon in Chapter 5 which explores the gendered material features and relational capacities of voices in academia. Chapter 8, titled "Cow Choirs," intertwines several feminist philosophical and ethical takes on the agency of non-human animals and matter, for example, the work of Rosi Braidotti, with voice studies and musicological perspectives within its approach to practices of Scandinavian vocal herding musicking.

Some of the essays, like Chapter 9, "Speaking of Atmospheres," investigate how voicings co-constitute the material and semiotic traits of specific sociopolitical practices, such as courtroom sessions, and also how non-human beings, for instance worms, can be understood as giving voice that calls out to other beings. The latter notion provides a valuable lesson about the existential companionship of humans and other species. Chapter 9 thus observes the workings of voice in more-than-human sociopolitical realities while doing so by engaging with select sound and media art projects. A similar focus on artistic voice practices and the rewards of studying the arts through new materialisms saturate many of the other chapters. This is true of Chapter 2, a powerful reconceptualization of musicality as emergent ways of doing produced by the intra-actions of vocal, musico-material practices. In Chapter 6, readers are invited to follow the traces of a site-situated performance from 2018 which investigates the forces of vibrational vitality emerging between different sorts of voices, including those of a mouth, a copper wire, and a museum car park, all staged through a new materialist artistic practice and an exercise in performance writing. Chapter 7 is, in turn, a rich meditation on an artistic practice co-created with the theremin.

Expanding the relations of new materialisms: Artistic research and somaesthetics

One more important characteristic of new materialisms, which, to be sure, characterizes many other fields as well, is their overlapping or entanglement with several adjacent theoretical projects and sets of approaches. Key cross-disciplinary areas of research adjacent to new materialisms include affect studies, which explores from multiple perspectives the capacities of human and non-human beings to affect one another (Gregg and Seigworth 2010; Seigworth and Pedwell 2023) and posthuman(ist) scholarship, which maps the fundamental coexistence of humans with the more-than-human world and the ethico-political stakes ensuing from this (Górska 2016; Bentivegna and Edlund 2022; Wilmer and Žukauskaitė 2023). Also, philosophical theories concerned with the processual, relational nature of reality, even when their developers do not espouse the term new materialism (Manning and Massumi 2014; Gatt 2017; Ingold 2022), as well as ecocriticism and the environmental humanities, share some of the terrain with new materialisms. The commingling of new materialisms with these other approaches can be discerned in different variations in this volume's essays. While commenting on this further in the later sections that introduce the book's three parts, here we already want to highlight two areas whose entanglement with both new materialisms and voice studies is a distinctive feature of this book: artistic research and somaesthetics.

As a bridge to that discussion, let us return to the crossroads evoked in the beginning of this introduction. Let us tune into the sound of three scholars intra-acting as a way of attentively allowing new materialisms to playfully be re-configured into new connections, stories and becomings. Let's re-begin like this:

> We walk through the village. Along the river. Air is moist, and trees are carrying thousands of crystal-clear water drops. Clouds above our heads are dark and voluminous. Ready to let go of more rain. Small pools have been formed on the gravel road underneath our feet. We breathe in the humid air. Eyes and ears are attending to the situation. Lungs filling. With freshness and life. Water seems to be present to all senses, when the sound of pushing water comes closer. Water is being held back. Collected. Before finding an opening in the resisting wall. The force creates an overwhelming power, rushing into the open air and further in to the river. It is as if nature allows its voice to matter. To make itself heard. Flowing, floating and flashing forth. Connecting to memories, to any previous experience of moving. We all have different images inside, while moving together for a few days along the same river. What we are about to share is awakened by our co-presence with/in the situation. And this sharing entails a sensitive attending to where we stand, what we are about to do, and how our actions will be collected and expressed. All that is encountered moves along in the process: the place, the atmosphere, our bodies functioning as part of what we know and what we have yet to discover.

> Being attentive. Opening the self along with the spacetime of a here and now. Allowing for the next scene to begin. Preparing the stage for a voice to appear. And merge with all that moves and takes place. Attentiveness makes space for all those questions that normally would be taken for granted. But in the vulnerable moment of bursting open, no question is irrelevant. All may care-fully be asked and re-asked. Opened and re-opened. The act of searching has no limit. Just like the river in flow. Just like a voice nurtured by breath and by air. Sensitivity to what is possible is about acknowledging both systematic organizing and poetic bewildering. A paradox in its becoming. And voice is the home of it all. As much as voice is the home of Nothing.

From here, writing is reconfigured and led through a conversation between three voice scholars from different backgrounds. Voicing is our common passion, as is also a love of learning and building knowledge beyond what is expected and foreknown. We needed to share our stories with one another in order to write together. Our words capture our lived experiences of the fields where we feel at home. Stories we can share between us. Stories that become more stories about more fields and more voice research, as readers will become part of this volume. Stories will then keep generating new understandings and voicings as far as anyone can imagine. All that is needed are crossroads where things begin again.

One such open-ended crossroads involves new materialisms and artistic research. As such, artistic research is not a new or exclusively recent practice. Learning to do, to make, to construct, and to perform in any possible way has always demanded specific modes of practicing, and time for reflection. This is a valid fact for any field of research. But artistic research has not always been included as an academic field equal to those representing "scientific" research – not until the beginning of the 2000s when some Faculties of Art obtained examination rights at doctoral level for artistic practice. This new academic direction has allowed artist-scholars to develop an infinity of alternative research strategies and methods – as "research results" – based on and led through subjective experiences in relation to artistic processing and meaning-making.

Artists began to explore different ways of shaping knowledge through any expressive format; new academic journals (such as *JAR – The Journal of Artistic Research*, *Ruukku – Studies in Artistic Research and VIS – Nordic Journal for Artistic Research*) were launched offering alternative opportunities for publishing peer-reviewed papers as "expositions," video and audio essays. The Research Catalogue (https://www.researchcatalogue.net/) was established by the newly founded Society for Artistic Research as an international database for this field. Within less than 25 years, the field has expanded rapidly. One of its strengths is that written language, as well as other forms of performative and poetic expression, have become equally important and accepted as means for transdisciplinary sharing and collaborating. In 2008, Henk Borgdorff wrote a key paper where he points out that artistic research "continuously starts anew" (2008: 95), and "creates room for that which is unthought, that which is unexpected: the idea that all things could be different" (Borgdorff 2008: 96). In his "A Manifesto for Performative Research," Brad Haseman (2006) notes, in turn, that performative research strategies ought to be considered next to quantitative and qualitative ones, while highlighting that the "principal distinction" of performative strategies "is found in the way it chooses to express itself" (Haseman 2006: 5). A text could be "any object or discourse whose function is communicative" (Haseman 2006). This is where new materialist theories and methodologies "flash up" (Barad 2017) into the field of artistic research – and especially into artistic voice research where psycho-physical sensations such as wonder and newness are directly experienced within the voice-practitioners' selves – thus providing helpful tools for communicating sensuous knowledge, linking poetic and scientific expression in the broadest sense – not in order to turn one research strategy or tradition comparatively against the other, but to allow for previously unknown ways of researching in order to "create a non-hierarchical network of irreducible difference" (Merten 2021: 5).

Another entanglement quite new to voice scholarship that is probed in this book emerges between new materialisms and pragmatist somaesthetics, an approach first devised by philosopher Richard Shusterman. In somaesthetics, embodied and esthetic phenomena are analyzed in cultural and political terms, embodied methods and practices are critically examined and developed, and embodied experiences are engaged in practice (Shusterman 1999, 2008, 2012). Even if Shusterman's thinking is not explicitly oriented toward new materialisms or posthumanist thought, other researchers have made connections between these areas and his work (Abrams 2004, 2012; Małecki and Schleusener 2015). Moreover, lines of connection have been drawn between new materialisms and John Dewey's pragmatism, in which Shusterman's somaesthetics is deeply rooted (Guernsey 2017). In particular, Dewey's and Arthur F. Bentley's pragmatist concept of *transaction* and Barad's new materialist concept of *intra-action*, both inspired by the thinking of physicist Niels Bohr, have been considered in parallel (Dewey and Bentley 1949; Hammarström 2010; Rosiek 2013, 2018; Fischer 2018). Chapter 4 of this book presents a new merger of a somaesthetic framework, manifest especially through the concept of transaction, and new materialist insights while examining the embodied experiences of singers with acid reflux.

Engaging with critiques of new materialist thinking

New materialisms have not gone unproblematized during the past 10–15 years. They have provoked criticism from a number of directions, and critical appraisals continue to appear in various humanities and social scientific publications. One key target of criticism already discussed above is how advocates of new materialisms have arguably overstated the novelty of their initiatives while downplaying the previous theorizations of materiality in Marxist and feminist research, for instance (Ahmed 2008; Hemmings 2011). Several articulations of new materialisms in the late 2000s and early 2010s did rely on rhetoric that differentiates new materialist stances from social constructionist ones, for example. However, more recent studies have emphasized the necessity of converging these sets of approaches in different fields (Lorenz-Meyer et al. 2019; Kontturi et al. 2025). We believe this latter ethos is at play in *Mattering Voices*, as the authors intertwine new materialist modes of thinking and feeling with diverse subject matters and research areas instead of working by way of antagonism or opposition.

Many chapters of the volume also speak back to another criticism of new materialisms, according to which attention to agential materialities, including non-human, risks sidelining the role of human sociopolitical practices and power dynamics in constituting reality (Hemmings 2011; Irni 2013). We believe the essays of *Mattering Voices* demonstrate the benefits of exploring material processes *as part* of cultural and sociopolitical analysis of voicings, without relegating questions of human agency or social organization to the margins (see Chapters 1, 3, 5, 6, and 8).

Yet another important criticism of new materialisms has concerned their whiteness and Western-centeredness in the sense that new materialist thinkers' reorientations toward matter and the more-than-human world have arguably shown little acknowledgment of the significance of materiality and non-human nature in many Indigenous bodies of knowledge (Sullivan 2011; Todd 2016). This astute critique is increasingly being addressed in new materialist studies, and it has begun to reshape their theoretical underpinnings and methodological practices. In this regard, Virginie Magnat does a wonderful job in her *The Performative Power of Vocality* (2019) in terms of investigating culturally diverse and Indigenous ontologies, epistemologies, and ethics as indispensable sources for voice scholarship and new materialist thinking alike.

We were determined to include culturally diverse contributions in *Mattering Voices*, as well, while critically acknowledging the still strong white dominance in academia. This aim was qualified and, to some extent, restricted by the book's purpose of showcasing voice research conducted with new materialisms, which remains a relatively novel set of approaches in the field of transdisciplinary voice studies. While guided by this purpose, we sought to invite contributors from diverse backgrounds and will admit that our efforts were not always successful, as many inspiring scholars had to decline the invitation because of their other commitments. Although the resulting cultural diversity of *Mattering Voices* is not as rich as it could be, the book nonetheless contributes to research that contests Euro- or Western-centered models of thinking and the prioritization of (particular kinds of) white humans. By working with new materialist feminist, environmental, postcolonial, process philosophical and also Indigenous approaches to materiality, the volume's chapters highlight such intra-active relationships between humans and non-humans, or individuals, materialities and environments, which powerfully challenge the atomistic idea(l)s of human agency inherent in Western, white traditions of thought.

Many chapters also engage with socially marginalized vocalizers, such as people with disabilities (Chapter 4) and transgender singers (Chapter 3), thus stressing their importance as voicers and research participants.

The question of critique is a crucial one never to be left aside, and therefore we want to return to one of the main ideas characterizing new materialisms, namely the "more-than" aspect. As Barad emphasizes in her introduction to *Meeting the Universe Halfway*, this aspect is about providing "an understanding of the role of human *and* non-human, material *and* discursive, and natural *and* cultural factors in scientific and other socio-material practices, thereby moving such considerations beyond the well-worn debates that pit constructivism against realism, agency against structure, and idealism against materialism" (Barad 2007: 26). The new materialisms emerging in *Mattering Voices* also participate in proposing "a rethinking of fundamental concepts that support such binary thinking, including the notions of matter, discourse, agency, power, identity, embodiment, objectivity, space, and time" (Barad 2007).

Doing voice studies through new materialisms

This last section of the introduction fleshes out how the ways of doing voice studies through the new materialisms that are presented in *Mattering Voices* seek to expand the understanding of three distinguishable but richly interrelated themes: ontologies of voice, vocal agencies, and the emergence and work of voicings in more-than-human realities. The focus on these themes is reflected in the book's tripartite structure, which we discuss next by illuminating the themes at the crossroads of transdisciplinary voice studies and new materialist theoretical currents.

Part I: Voicing ontologies

... wondering as voicing ...

Voicing can be understood as a way of letting go; as a way of dropping the jaw, opening the self as an instrument, learning processes of controlling and releasing – through every breath, every sound – to let air enter through the nose and the mouth; to fill the lungs as well as any imagined space of the living, breathing body. It is about conscious control as much as about remaining in a state of dynamic and fluid *wondering*. A voice often aims at touching the "listener" (here meaning the one who is opening all senses), creating a presence; while at the same time, a voice cannot be physically touched in the same palpable sense as other body parts. Voice is about breathing and energies, about learning to freely make sense, while simultaneously a voice cannot always be explained, but it can still be experienced; lived through and sensed as nothing more than nonsense. As such, voicing is very much a paradox. Voice was even identified as a "Nothing" when Western public opera was established as a genre in Venice around 1637 (Calcagno 2003). And it is through Nothing that a significant relation between voicing and new materialist terminologies can be invoked. Nothing, as well as voicing, can be endlessly reconfigured. Karen Barad writes: "Shall we utter some words about nothingness? [...] Perhaps we should let the emptiness speak for itself" (2012: 4). And suddenly we hear the sound of

... voicing as nothing ...

returning, as sounding paradoxes of both emptiness and fullness, as was already confirmed by the *Accademia degl'Incogniti* who produced the first operatic events together with composers such as Monteverdi and Sacrati in 17th-century Venice (Belgrano 2011, 2019). At that time, Nothing was considered both an object to be heard and a practice performed by singing actors. Barad continues: "Nothingness is not absence, but the infinite plenitude of openness. [...] if we listen carefully, we can hear the whispered murmurings of infinity immanent in even the smallest details. Infinity is the ongoing material reconfiguring of nothingness [...]. Infinity and nothingness are not the termination points defining a line. Infinity and nothingness are infinitely threaded through one another, so that every infinitesimal bit of one always already contains the other" (2012: 16–17).

... nothing as phenomena ...

In a short reflection dedicated to "air," Caroline Gatt invites the reader into a "dance" through fragments, words and thoughts beginning with "Becomes Breath, Becomes Song, Becomes Air," while ending with "Song, voiced breathy air, is a way, a path, along which we can encounter our emergent constitutive histories" (2017: 5–8). In between these passages she introduces a range of characters, wisdom, touches, facts, signs and codes. Beginning and end, both becoming much more than what they first seem to be. Gatt invites readers into a world as "an open process of mattering through which mattering itself acquires meaning and form through the realization of different agential possibilities" (Barad 2007: 141). Gatt's reflection thus becomes an "ongoing material reconfiguring" (see above) of air; more than a text, a set of words, a trace reflecting the past. Her words become "relations of exteriority, connectivity and exclusion" reconfigured as "phenomena – dynamic topological reconfigurings/entanglements/relationalities/(re)articulations of the world" (Barad 2007; see also Barad 2017). Moreover, the narrative presented by Gatt draws even closer to the readers when they get out of their chairs, moving toward bookshelves searching, for example, for *Resonance* (Rosa 2019) as another collection of words about breathing and voicing; where readers find that "certainty and rhythm of breathing are parts of the realm of ontological security [...]. Hence [...] when ontological certainties become uncertain, say, because we suddenly find ourselves in mortal danger; or in those moments [...] when the ground begins to move beneath our feet, we say that we are *short of breath*, or even *left breathless*. [...] Grasping or holding one's breath in sympathy [...] is here itself the expression of an experience of social resonance" (Rosa 2019: 53–54). Readers may look out through windows, holding on to breath while considering the moment of *momentum,* extended tension in time enabling a conscious encounter with "an ongoing reworking of the notion of dynamics itself. Dynamics [as] matter not merely of properties changing in time but of what matters in the ongoing materializing of different spacetime topologies. The world is intra-activity in its differential mattering." And readers may allow for a phenomenal "intra-activity in its becoming" (Barad 2007: 141) to keep sounding

... phenomena as "intra-activity in its becoming" ...

Such is the stage for the first two chapters of this volume. With Jannie Pranger's detailed engagement with Karen Barad's concepts in Chapter 2, one can learn how to relate to Baradian new materialist terminology through a vocal performing scholar's way of intra-active reading. Chapter 1, co-written by Laasonen Belgrano and Price, presents, in turn, a spacetime of an operatic "mad scene" based on a diffractive contemporary dialogue between academic – intra-active – thinking and the art of vocality in action, very much inspired by similar dialogues performed in 17th-century Venice in relation to the birth of public music drama. Both chapters call for thinking/voicing/acting as ways of moving and entangling. Overflowing. Spilling over. *More-than* as an alternative to *either or*. Chapter 2 searches for ways to articulate and explain how voicing can be understood as a praxical means for "Meeting the Universe Halfway" (Barad 2007), whereas Chapter 1 allows for voicing to praxically lead any agents (authors as well as readers) through an open process – the phenomenon of voicing-as-thinking-as-nothing-and-more. This processing of words and making of meanings as intra-activity could then spill over into any of the other chapters, and even further into the everyday living of anybody who encounters the narratives of this book. In any possible way, performing an

... "intra-activity in its becoming" as slow diffractive reading ...

which stands as an "apt metaphor for describing the methodological approach [...] of reading insights through one another in attending to and responding to the details and specificities of relations of difference and how they matter" (Barad 2007: 71). This methodology has recently been explored in *Diffractive Reading. New Materialism, Theory, Critique* (2021), an anthology edited by Kay Merten, where he points out that any two ways of doing or arguing for something "become entangled like two waves in diffractive phenomena," thus creating "a non-hierarchical network of irreducible differences" which reveals "texts as always already entangled with one another, without any priorities or hierarchies" (Merten 2021: 5–6).

Adopting this understanding, *Mattering Voices* also proposes ways of "drawing attention to an important point about the need to blast open terms such as" (Barad 2017: 65) voice, voicing, and vocal praxis as a way of performing, led through somatic-philosophical concepts and narratives (Norbakken 2024). Engaging intra-actively with this volume hopefully encourages modes of "slow reading – an arresting of thinking, at least a slowing down … a practice of opening up to the possibilities of important insights flashing up […] an anticapitalist praxis" (Barad 2017). Barad describes this method as one concerned with encountering the nature of matter, which this volume translates into ways of encountering the nature of mattering voices.

… slow diffractive reading as voicing …

Following this proposed new materialist methodology is not at all that far from practices of voicing. The beauty is that the crossroads of wonder hopefully invoked by this book act as an invitation to breathe, to sing, to speak, to think and write, to let ourselves move in any possible direction as scholars-actors-voicers-beings-mattering voices. What this book offers are fragments of texts among many other available narratives in the broadest sense, and *Mattering Voices* could never claim to be anything else than a fragment in itself. Yet, this does not preclude the wish that this collection – in its limitedness – will begin right here and now, just as when we open our eyes and bodies-as-somatic-systems, and let our lungs and minds be filled with wonder and curiosity for all that has not yet been read, described, and understood, in any spacetime imaginable. At this crossroads – being the very text you have begun to read – words, concepts and narratives will appear, dealing in a vast number of ways with how theory and practice may (be)come together through

… voicing ontologies as ways of wondering …

Beginning. In the moment one becomes aware. This is where voice and new materialisms coincide. Diffract. And may move into unknown spacetimes. Awareness springs out of this moment. Vibrating. Connecting knowing and feeling. Expanding. Voice is flashing open, as thinking and sensing, allowing unreason to make some kind of ontological sense, perhaps unexpected. A sense-making that is most probably different in and for everyone. In an infinite number of manners, the crossroads of voice studies and new materialisms are constructed by paradoxes and tensions. *Both* of letting go *and* of controlling. Like when a voice starts to disperse into spacetime, allowed to be free, "let out" and expressed, while also being carefully, systematically and consciously controlled. All performed as a way of wondering.

Part II: Voicing intra-active human agencies

A conventional way to understand a voice is that it is located in a human subject while belonging to that subject and emanating from their specific body as a product of their intention and action. According to this notion, voice articulates and represents human experience, expresses the voicer's individual and/or socially categorizable identity, and is ontologically a relatively stable entity (Eidsheim 2019: 1–4). Although the need to consider voices instead as changeable and fundamentally shared phenomena has been widely explored in voice research (Cavarero 2005; Neumark 2010, 2017; Eidsheim 2019), there is still a tendency to perceive the voice primarily as an instrument of language or music. A new materialist framework provides effective ways to unsettle these conventional truths. New materialist perspectives encourage a rethinking not only of voice but also such associated phenomena as human subjectivity and individuality, agency, experience, and the body. For example, Lisa Mazzei and Alecia Jackson discuss the possibilities of understanding the voice as more than just a phenomenon tethered to language or as sounds emanating from the human body (Mazzei and Jackson 2017: 1090). Rachelle Chadwick proposes, for their part, that researchers shift their focus from seemingly uniform and easily classifiable voices to voices' elusiveness (Chadwick 2021: 79).

The contributors to *Mattering Voices* address voicing events and practices through new materialist perspectives that shed some novel light on voices' relationality and interconnectedness, vital materiality, actuality and tangibility, sensuality, and in-betweenness. The three chapters of the book's second part are connected by their focus on voicing intra-active human agencies, or how such agencies come about during processes of voicing. These essays explore how vocal events, actors and characteristics are constituted as acceptable or unacceptable, appropriate or inappropriate within the dynamic operations of various material, social, cultural, and esthetic intra-active entanglements.

In Chapter 3, Milla Tiainen explores, in collaborative contact with Finnish singer and transman Demian Seesjärvi, the relationships of voice, the voicer's gendered senses of self, and the environments of their voicings as constantly emergent and co-constitutive. She examines what this relationality might mean to a transmasculine vocalizer in particular. The chapter also explores the agential role of materialities within the relations at stake, which range from the hormonally and singing-technically transforming features of Seesjärvi's voice and body to the cultural milieus in which he has sung and the more-than-human material entities, such as sound technologies, which provide him access to diverse vocal styles. Tiainen analyzes the environments of voicings as multidimensional socio-material phenomena. She demonstrates how the relations among voice, voicer, and the given surroundings have been constitutive of Seesjärvi's selfhood, both undermining and affirming his transmasculine senses of self. On a theoretical and methodological level, the chapter shows how transdisciplinary voice studies, research on transgender voices, and new materialisms can mutually enrich each other while jointly advancing scholarship and awareness on the importance of voicings to transpeople.

Chapter 4 by Anne Tarvainen discusses experiences of the body's materiality among singers with acid reflux while drawing on Richard Shusterman's somaesthetics and Bentley and Dewey's concept of *transaction*. Tarvainen demonstrates how reflux-induced vocal problems generate perceptions of a fragmented body unable to express or experience esthetically, whereas in the best singing experiences the body emerges as a whole, and stimulated by the vibrations of sound, may even feel like losing its boundaries. Bringing new materialist concepts into intra-action with those of pragmatist somaesthetics, Tarvainen suggests these feelings of relatedness and unity could be conceptualized as *transactional experiences* of voicing. In Chapter 5, Monika Rogowska-Sangret and Malou Juelskjaer address wounded voices as a trans-corporeal phenomenon in voicing situations that take place in contemporary academia. They theorize voice as touch and response-ability while presenting and analyzing their personal experiences as academic teachers and researchers who have attended voice coaching sessions. Ultimately, Rogowska-Sangret and Juelskjaer explore how voices, bodies and relational worlds might become different – less hierarchical and more response-able – in contemporary neoliberal academia.

As exemplified by the chapters just introduced, new materialist approaches propose that voice is created in varying assemblages of multiple agents. Thus, a voice is inherently *relational*. It does not wait within an individual body-mind to be expressed, and nor is it a straightforward result of human intentionality (Chadwick 2021: 79; Mazzei and Jackson 2017: 1090; cf. Eidsheim 2019: 9–14). Chadwick points out that vocal utterances always emerge in relation to other voices, material realities, and intersubjective encounters. In this sense, the voices emitted in such encounters do not "belong to us," the vocalizers, because they are produced in complex socio-material, embodied, and discursive engagements. As a result, normative assumptions about voice as a source of authenticity, liberation, or empowerment come under challenge (Chadwick 2021: 97). The notion of "one's own voice" as something uniquely individual and independent of the given situations is also reconfigured (Magnat 2019: xii).

These insights give reason to ask whether the embodied movements and aerial vibrations of different kinds of voicings can really be considered "properties" or "characteristics" of the voicer. Ways of singing, for instance, are largely based on cultural habits, esthetically desired vocal actions, and learned embodied repetitions that are not just attributes of an individual singer but permeate countless singing bodies, albeit variably. Simultaneously, the ongoing relationality of vocal expression has the potential to produce something unexpected. Tiainen has conceptualized this as "reconfiguring relationships," whereby a singer discovers a new sense of bodily engagement with vocal expressions that are unconventional within the given vocal tradition (Tiainen 2013: 399).

Just as new materialist theories encourage an understanding of (human) agency as a potential arising from within intra-acting elements, voices, too, can be thought of as a collective output of (partially) material assemblages. Mazzei and Jackson suggest accordingly that it is fruitful to direct attention to the movement and the agential force of voices in different configurations (Mazzei and Jackson 2017: 1093). Also, since voice is not something that belongs to someone, it is untenable to draw a strict dividing line between things that have a voice and those that do not (Mazzei and Jackson 2017: 109; see also Neumark 2017 and Preston's chapter in this book).

New materialist approaches allow for exploring voices as action rather than mainly as representation, while offering ways to consider them as ever-morphing, re/active, situational processes. This entails recognizing the agency and vitality of voice-related *materialities* – both those of human bodies and the other material forces and substances contributing to vocal events. More than a representation of a preexisting experience, a voice is an agent that acts in the present within the given assemblage (Mazzei and Jackson 2017: 1092). Thus, voices can be rethought as "voicings": emergent relationalities and processes (Chadwick 2021: 97; see also Tuana 2008: 189; Fast and Tiainen 2018). This presents a departure from voice-related essentialism and from the humanist subject as the assumed primary source of experience, knowledge, and truth (Mazzei and Jackson 2017: 1091).

The chapters in the second part of this book, but also elsewhere, explore how liveness, dynamic forces, intensities, movements, vibrations, resonances, and affects shape and are shaped by voicings. Voicings are vibrant, and they act with greater or subtler force in assemblages of bodies, words, histories, and more (Mazzei and Jackson 2017: 1092). In new materialist theories, both human and non-human entities are seen as mutually affective actors (Bennett 2010: 20–38; Tiainen 2013; Mazzei and Jackson 2017: 1092; see also the next section). In this respect, agency is always distributed and shared, which leads to a notion of voices as fundamentally interconnected or intra-actively formed. Voices transform, respond to, and cause reactions within intra-actions of material, symbolic, social, mental, and other elements. "[Voice] is always in response, in relation, and characterized by a jostling movement of multiple imagined, real, moral, and fleshy voices" (Chadwick 2021: 98).

Once they sound out, voicings, although processual, are also actual and tangible; embroiled in specific spaces, locations and situations. As affective and response-inducing, they relate to the senses and sensory experience, which, like agency and voice, can be understood as emerging in wide tangles of matter exceeding the individual human body. Although the concept of experience has not been theorized in depth in many new materialist (or posthumanist) texts, it is frequently touched upon in new materialist writings on voice (Tiainen 2007, 2013; Fast 2008, 2016; Fast and Tiainen 2018; Leppänen and Tiainen 2018; Laasonen Belgrano 2015, 2018, 2019, 2021, 2025; Norbakken 2024). Neumark (2010: xvi) discusses the inextricable link between the esthetics of voice and sensory experiences of the whole body, created jointly by the senses. Magnat (2019: 130) describes singing as a carnal act where a singer senses the acoustic properties and proprioceptive sensations of their voice in their body. In the chapters of *Mattering Voices*, the embodiment and sensuality of voicings are explored with such notions as sensibility, intimacy, breath, touching and being touched, affective attunement, and empathy. Human bodies-subjects always intertwine with the world beyond their skins (cf. Sullivan 2001). New materialist(-inspired) concepts like trans-corporeality, the in-between, liminality, trans-materiality, and the more-than-human effectively illuminate this relationality, as the chapters of this book demonstrate.

New materialist strands of thought also propose some new understandings of the nature and frontiers of experience. The ability to feel is not only a human property but may also be a potential of non-human, even inorganic, matter (Shaviro 2014, according to Parikka 2015: 69). As Barad has argued, "[m]atter feels, converses, suffers, desires, yearns and remembers" (Dolphijn and van der Tuin 2012: 59). Magnat stresses the need for a poetic sensibility that may help voice researchers understand the complexity of the world better than the traditional discourses of Western science and philosophy (Magnat 2019: 162). This brings us to the poetic modes of writing that a number of scholars and artist-scholars, including several contributors to this book, practice when doing new materialist research. These ways of writing enable fresh perceptions and sensibilities regarding the entanglement of bodies, voices and other materialities. Arguably, modes of voicing, from singing to speaking and beyond, are particularly predisposed to producing experiences of connection: of coming together, being part of larger assemblages, and opening beyond one's apparent material boundaries. This may be one of the key reasons why new materialist approaches resonate so richly with vocal experiences – and work productively in voice studies.

Despite unfolding as actual sounds, the relationality of voice makes it often impossible to locate firmly, as it remains somewhere *in-between* (Macpherson and Thomaidis 2015, introduction; Laasonen Belgrano and Price 2025b). The vibrations of voicings quite literally dissolve into thin air, as has been repeatedly discussed in sound and media research (Tiainen 2013: 393). Ben Macpherson has proposed that when a vocal sound has departed the body and is not yet accessible to listeners, we can speak of a sonic in-between. It is in this space that the shared experience of the producers and recipients of voice takes shape (Macpherson 2021: 221). Chadwick, too, discusses the in-betweenness of voices, examining how they are alive and how meaning flows between the uttered words in the form of breaths, rhythms, laughs, pauses, and intonations (Chadwick 2021: 80). They also point out that voice is a kind of paradox (cf. Belgrano 2016), a borderline concept and phenomenon (Norbakken 2024, being at once language, embodiment, speech, silence, presence, and absence – yet without belonging neatly to any of these dimensions (Neumark 2010; Chadwick 2021: 77).

To date, a large part of voice research has been conducted in the medical framework, where the human body is understood as a physiological entity divided into functional units like organs and muscles. Arguably, these studies tend to produce knowledge about "a general body" that belongs to no one and occupies no specific situations (Cavarero 2005). Within the widening field of transdisciplinary voice studies, approaches that diversify this "singular" truth about the vocal body have been developed (Thomaidis 2019b: 300; Kapadocha 2021b: 1). This is particularly important given that many vocal pedagogies are based on research which thus influences how voice is taught and enacted in daily life. Experimental, cross-cultural, and non-Western approaches to vocality are too often overshadowed by the systematic and institutionalized practice of voice (Magnat 2019: xiv). Many Western singing cultures remain exclusive, with limited places available for only the select few (Chapter 3 in this volume). The new materialist notion of intra-action elaborated across this book encourages one to explore how voicings bring humans – and other beings – together, creating meaningful places in the world, communities, and (sub)cultures for those involved.

As illustrated by the example of balcony singing with which we began this introduction, many humans (and non-humans) yearn for connection and safety. According to Stephen Porges, the developer of the polyvagal theory, this tendency dwells in the very fabric of human bodies, affects, nervous systems, and vocalizations (Porges 2011: 282–283, 324–325). In the chapters of this book, the interconnective nature and capacities of voice are highlighted with concepts such as relationality, dis/connection, communication, encounters, caring, entanglements, intra-action, trans-action, assemblages, co-constitution, and response-ability (see, e.g., Pranger's chapter). Vocal activities involving so-called others beyond the normative Western subject, be these non-human beings or marginalized humans, foreground the ethically important issues of ability and disability, plurality and inclusivity (e.g., Nauha's, Preston's, and Tiderman-Österberg's chapters in this volume). New materialist theories are specifically attuned to the relation-making and category-crossing agencies of voicings. At the level of lived praxis, collective encounters, where diverse voices and bodies cooperate without performance pressure, exploring new vocal territories or connecting to ancient rituals, have the potential to question received, limiting Western truths about voice (see, e.g., Magnat 2019: xv, 180–183; Fast 2008, 2016).

Part III: Voicing the more-than-human

The assemblages of voicing studied in this book practically always involve non-human materialities and beings in addition to the human ones. In the chapters, these materialities range from rocks, bodies of water and soil, and natural acoustic sites to non-human animals such as cows, goats, or worms. They also include sound production, modulation and dissemination technologies. To echo voice studies scholars Francesco Bentivegna and Sophia Edlund, as well as the new materialist notion of the agential capacities of all matter discussed earlier, the contributions to *Mattering Voices* do not approach the non-human participants of voicing events as a subordinate and passive opposite of vocalizing humans. Instead, the human and the non-human fundamentally coexist and depend on each other for their existence – both in general ontological terms and in highly situated ways. They co-create activities, such as potentialities, processes, and experiences of voicing (Bentivegna and Edlund 2022: 3).

Theoretically, the notion of co-implicated human and non-human ways of being highlights a specific overlap: that between new materialisms and *posthumanist* strands of thought. This overlap is evident in several voice studies publications thematically close to *Mattering Voices*. For example, in their editorial to the special issue on "Posthuman Voices" in *Journal of Interdisciplinary Voice Studies*, Bentivegna and Edlund (2022: 3) suggest that the discussion on posthumanism and voice has been initiated by scholars and practitioners in several fields, one of which is new materialisms. In the introduction to her book *Voicetracks*, Norie Neumark proposes a similar relationship between posthumanism and new materialisms. She maintains that posthumanism refers to an era characterized by a growing need to find alternatives for anthropocentric ethics, politics and esthetics, whereas new materialisms represent one group of theories that outline such alternatives (Neumark 2017: 2, 171 n1).

The overlap between posthumanism and new materialisms also applies to the work of several theorists that have been highly influential within research associated with either one or both labels. For instance, feminist philosopher Rosi Braidotti was among the first to coin the concept of new or neomaterialism in the early 2000s, and she is widely considered a central initiator of these perspectives (Braidotti 2002; Dolphijn and van der Tuin 2012: 19–37). Simultaneously, she has played a prominent role in advancing the concepts of posthuman and posthumanism (Braidotti 2013; Braidotti and Hlavajova 2018). Likewise, a combined focus on materiality and posthumanist perspectives imbues the work of Karen Barad, whom we have repeatedly cited, since their early publications on these topics, as exemplified by the title of their article from 2003, "Posthumanist Performativity: Toward an Understanding of How Matter Comes to Matter" (Barad 2003).

Despite the overlaps, there are features that can be seen to distinguish new materialist and posthumanist research fields from one another. Also, as editors of this book on voice studies and new materialisms, we do not fully share the view that new materialisms would be a subgroup of a broader range of posthumanist theories. Considered as diverse fields, both new materialist and posthumanist approaches are linked to theoretical strands that pull in mutually incompatible directions. In the case of posthumanism, these strands include transhumanist stances whose concern with the enhancement of human capabilities through genetic engineering, AI and other technologies is a far cry from the efforts to dethrone anthropocentrism by critical posthumanists such as Braidotti and many others. As discussed earlier, in the case of new materialisms, the insights of OOO, with its emphasis on a self-directed material world out of reach for human awareness, diverge from the notion of co-constitutive relations that underlies most variants of new materialist and also posthumanist thinking.

There are, then, differences between and within posthumanist and new materialist studies. To simplify, critical posthumanist approaches, which are being elaborated in voice studies, including this volume, focus primarily on interrogating and unraveling the Eurocentric, humanist concept of the human subject as the measure and ruler of all things. New materialist theories are, in turn, principally concerned with developing ontologically dynamic understandings of organic and inorganic matter, while drawing on diverse bodies of thinking such as process and eco-philosophies, feminist theories, complexity, chaos and other scientific theories, and phenomenology. We thus consider it important to acknowledge the specificities of new materialist and posthumanist research assemblages without just collapsing them into each other. At the same time, these areas engage with similar questions in closely interconnected ways. For example, the posthumanist understandings of human activities as co-created by non-human beings are crucially supported by how new materialisms expand the scope of agency to an array of material beings and processes.

In this volume, the adaptations of new materialist theories contribute to an understanding of voicing activities and voicers as products of and actors in relational webs always wider than the human. These adaptations encourage further intra-action between new materialist and posthumanist perspectives. While the commingling of such perspectives characterizes many chapters of this book, it is perhaps most evident in the four contributions that constitute the volume's third part, "Voicing the more-than-human." The notion of more-than-human can be used interchangeably with that of non-human, which is also widely employed in new materialist and posthumanist research. Yet, by including the first notion in the title of the third part, we wish to emphasize the unfolding of the world's processes in vibrant *excess* of human ones, and the impossibility of neatly separating human characteristics and capacities from other modes of being – whether at the level of individual embodiment and voicing or larger practices. Thus, the concept of more-than-human does not signify the superiority of some species or entities (e.g., technologies) over humans, and nor does it connote fantasies of technologically or otherwise enhanced superhumanness. It refers to multitudes of organic and inorganic beings that have agential capacity alongside humans. The "more-than" also indicates that humans' interdependence with other beings may, and often does, lead to somewhat unpredictable changes in what it means to be human (Massumi 2013: xx–xxiii).

The section "Voicing the more-than-human" opens with Chapter 6, by Julieanna Preston, titled "Motor-mouthing." Preston extends the capacity to give voice – to make sounds that engage other beings and call forth meanings – from human fleshy bodies to things of various kinds, including mechanical fan motors and a nation's treasures held in a museum (cf. Neumark 2017). The chapter provides a first-hand exploration of the technical, cultural and theoretical forces that shaped *RPM hums*, a live art performance where Preston aimed to make the vitality of more-than-human material things palpable by absorbing and emulating their vibrations. Characterizing the performance and her related performance writing as attempts to practice a new materialist ethos, Preston investigates the activeness of materials and objects traditionally considered inert. As she puts it, the *RPM hums* performance entertained the proposition that entities such as fan motors might exhibit agency independent of human beings by prompting mutually affective encounters between different "tuning" bodies.

Chapter 7 by Tero Nauha resonates with Preston's proposal, as it explores the playing of the theremin on the basis of Nauha's performance-thinking practice co-created with this electronic instrument dating from the early 20th century. Rather than focusing on technical human virtuosity often associated with the theremin, Nauha rethinks theremin playing as an intra-action; not as a coming together of separate player and instrument, but as a mutual constitution, where the theremin, too, has a capacity to respond. Specifically, Nauha is interested in the theremin's singing voice. He argues this voice is less an imitation of a human voice than a performative posture of a non-human voice that was not possible before the theremin's invention and which comes into being in the creative intra-actions of human and more-than-human agency Nauha calls *thereminging*. Besides contributing to new materialist and posthumanist voice studies, Nauha's chapter demonstrates the affinity between new materialist theories, such as Barad's, and process philosophies ranging from the work of Baruch Spinoza and Bracha Ettinger to that of François Laruelle, from whom Nauha adopts the concept of non-standard.

Chapter 8, "Cow Choirs: Singing-with More-than-human Herds," by Jennie Tiderman-Österberg is a beautiful attempt to understand how voice is produced by more-than-human relations, or how humans vocalize-*with* other material creatures. Tiderman-Österberg explores the living tradition of Scandinavian vocal herding musicking that involves long-distance calling named *kulning*, as well as songs and humming while milking and herding non-human animals. Her research grows from her own practicing of this tradition as a vocalist and music scholar in the context of Nordic *fäbod* or *seter* farming, where farmers and their companion species transfer to satellite farms and surrounding forest pastures for the summer. By thinking with new materialist, ecocritical, affect studies and posthumanist perspectives while focusing on human-cow relations, Tiderman-Österberg crafts a revived approach to vocal herding musicking by showing how it is about more-than-human intra-actions and affective transmissions. She proposes this multispecies practice as one micropolitical step toward sustainable futures and ethical relations that include the more-than-human.

In the book's final chapter (Chapter 9), "Speaking of atmospheres: More-than voice and voice of the more-than," Norie Neumark initiates a conversation between voice, place and new materialisms by exploring how atmospheres give voice. Expanding on her earlier pioneering work with new materialisms, Neumark argues that attention to the more-than-humanness of voices – of diverse things, from a stenographer to steel bullets and worms, shaping the acts and sounds of voicing – can enrich understandings of atmospheres and the related agencies of voice. To explore these matters, Neumark engages with several artworks concerned with the atmospheres of legal institutions, as well as with her own art project with non-human animals and how the emplacement of this project in different locations evokes different atmospheres. Neumark demonstrates that the focus of new materialist thinking on "material agencies and lively assemblages" foregrounds how voices co-compose atmospheres, how atmospheres voice themselves, and how various cultural and political forces saturate and are made possible by more-than-human voicings and atmospheres.

•••

At this point, there is only one further "more-than" that we, the editors of *Mattering Voices*, wish to invoke as regards the intra-active crossroads of voice studies and new materialisms: that is, the open-ended one created by the ways in which you, the readers, will engage with this collection, thereby letting it re-begin. It is our wish that the intra-actions of voice studies and new materialisms proposed throughout this volume will become indeterminably more-than any of the contributors could foresee when the chapters are read, felt and rethought through, challenged and pulled to varying relations and uses by voice researchers and artist-scholars and, hopefully, by interested folks in other fields. Thus, we embrace an onto-ethico-epistemological humility in the face of what voices, voice studies and new materialisms might be and do, and what they may still become.

Notes

1 This passage, along with the other later indented passages, was originally written by Elisabeth Laasonen Belgrano and then further sculpted by all the three authors of this introduction. The passages were inspired by our mutual discussions during a writing retreat that we undertook together in Fiskars Village, a well-known center of design and arts in Finland, when planning the introductory chapter at hand, and by our co-emergent sensory and affective experiences of the retreat's physical surroundings. Within this text, the passages are offered as an alternative way of approaching the book's subject matter and potentialities. Echoing modes of writing that can be found in artistic research – a field represented by several chapters of this volume in relation to voice and music – the passages deliberately differ in style from conventional scientific academic prose.

2 Following feminist theorist Nina Lykke's definition (2010: 27), we prefer the term transdisciplinary to that of interdisciplinary because it highlights how the expanding field of voice studies contributes to the exploration and reworking of such ontological, epistemological, and ethical questions that pertain across various disciplines, instead of being "owned" by any single area. Voice studies may also, for their part, reconfigure existing disciplinary boundaries and scholarly canons.

3 For a definition of the more-than-human, see section titled "Part III: Voicing the more-than-human" of this introduction.

References

Abrams, Jerold J. 2004. "Pragmatism, Artificial Intelligence, and Posthuman Bioethics: Shusterman, Rorty, Foucault." *Human Studies* 27 (3): 241–258. https://doi.org/10.1023/B:HUMA.0000042130.79208.c6

Abrams, Jerold J. 2012. "Shusterman and the Paradoxes of Superhuman Self-Styling." In *Shusterman's Pragmatism: Between Literature and Somaesthetics*, edited by Dorota Koczanowicz and Wojciech Małecki, 143–159. Amsterdam: Brill | Rodopi. https://doi.org/10.1163/9789401207638_011

Ahmed, Sara. 2008. "Imaginary Prohibitions: Some Preliminary Remarks on the Founding Gestures of the 'New Materialism'." *European Journal of Women's Studies* 15 (1): 23–39.

Alaimo, Stacy. 2008. "Trans-Corporeal Feminisms and the Ethical Space of Nature." In *Material Feminisms*, edited by Stacy Alaimo and Susan Hekman, 237–264. Bloomington: Indiana University Press.

Alaimo, Stacy. 2010. *Bodily Natures: Science, Environment, and the Material Self*. Bloomington: Indiana University Press.

Alaimo, Stacy. 2016. *Exposed: Environmental Politics and Pleasures in Posthuman Times*. Minneapolis: University of Minnesota Press.

Alaimo, Stacy, and Susan Hekman (Eds.). 2008. *Material Feminisms*. Bloomington: Indiana University Press.

Arlander, Annette. 2014. "From Interaction to Intra-Action in Performing Landscape." *Artnodes* 2014 (14). https://doi.org/10.7238/a.v0i14.2407

Åsberg, Cecilia, and Rosi Braidotti (Eds.). 2018. *A Feminist Companion to the Posthumanities*. New York: Springer.

Åsberg, Cecilia, Kathrin Thiele, and Iris van der Tuin. 2015. "Speculative before the Turn: Reintroducing Feminist Materialist Performativity." Edited Special Section "New Materialism." *Cultural Studies Review* 21 (2): 145–172.

Barad, Karen. 2003. "Posthumanist Performativity: Toward an Understanding of How Matter Comes to Matter." *Signs* 28 (3): 801–831. https://doi.org/10.1086/345321. Accessed 30 August 2023.

Barad, Karen 2007. *Meeting the Universe Halfway: Quantum Physics and the Entanglement of Matter and Meaning*. Duke University Press. *JSTOR*. https://doi.org/10.2307/j.ctv12101zq. Accessed 30 August 2023.

Barad, Karen. 2012. "On Touching—The Inhuman That Therefore I Am." *Differences* 23 (3): 206–223. https://doi.org/10.1215/10407391-1892943

Barad, Karen 2017. "What Flashes Up: Theological-Political-Scientific Fragments." In *Entangled Worlds: Religion, Science, and New Materialisms*, edited by Catherine Keller and Mary-Jane Rubenstein, 21–88. New York, USA: Fordham University Press. https://doi.org/10.1515/9780823276240-002

Barrett, Estelle, and Barbara Bolt (Eds.). 2013. *Carnal Knowledge: Towards a 'New Materialism' through the Arts*. New York: I.B. Tauris.

Barthes, Roland. 1977. "The Grain of the Voice." In *Image, Music, Text*. Essays selected and translated by Stephen Heath, 179–189. London: Fontana Paperbacks.

Belgrano, Elisabeth L. 2011. "Lasciatemi *morire*" *farò* "La Finta Pazza": Embodying Vocal Nothingness on Stage in 17th Century Italian and French Operatic Laments and Mad Scenes, ArtMonitor. Doctoral dissertation. University of Gothenburg, Gothenburg. http://konst.gu.se/english/ArtMonitor/dissertations/elisabeth-belgrano

Belgrano, Elisabeth L. 2015. "Ornamentation based upon More-Than-Human-References: Moving Towards on Ecology of Trust". Proceeding. CARPA4, Colloquium on Artistic Research in Performing Arts, The Impact of Performance as Research, Theatre Academy Helsinki, 11–13 June. http://nivel.teak.fi/carpa4/ornamentation-based-upon-more-than-human-references-moving-towards-an-ecology-of-trust-elisabeth-laasonen-belgrano/. Accessed 15 July 2025.

Belgrano, Elisabeth L. 2016. "Vocalizing Nothingness: (Re) Configuring Vocality Inside the Spacetime of Ottavia." *Journal of Interdisciplinary Voice Studies* 1 (2): 183–195. http://www.intellectbooks.co.uk/journals/view-Journal,id=248/

Bennett, Jane. 2010. *Vibrant Matter: A Political Ecology of Things*. Durham: Duke University Press.

Bentivegna, Francesco. 2019. Becoming Robot through Voice: Training in Artificial Voices. *Theatre, Dance and Performance Training* 10 (3): 439–447. https://doi.org/10.1080/19443927.2019.1634639

Bentivegna, Francesco, and Sophia Edlund. 2022. Issue "Posthuman Voices: Channels across Time and Shared Memories." *Journal of Interdisciplinary Voice Studies* 7 (1). https://doi.org/10.1386/jivs_00051_2

Bolt, Barbara. 2004. *Art beyond Representation: The Performative Power of the Image*. London: I.B. Tauris.

Borgdorff, Henk. 2008. "Artistic Research and Academia: An Uneasy Relationship." In *Autonomi och egenart: konstnärlig forskning söker identitet*, edited by T. Lind, 82–97. Stockholm: Vetenskapsrådet.

Bosma, Hanna. 2013. *The Electronic Cry: Voice and Gender in Electroacoustic Music*. PhD dissertation. University of Amsterdam.

Braidotti, Rosi. 2002. *Metamorphoses: Towards a Materialist Theory of Becoming*. Cambridge: Polity.

Braidotti, Rosi. 2013. *The Posthuman*. Cambridge: Polity.

Braidotti, Rosi, and Maria Hlavajova (Eds.). 2018. *Posthuman Glossary*. New York: Bloomsbury Academic.

Bryant, Levi R. 2011. *The Democracy of Objects*. London: Open Humanities Press.

Butler, Shane. 2021. "Animal Listening." *Journal of Interdisciplinary Voice Studies* 6 (1): 27–38. https://doi.org/10.1386/jivs_00035_1

Calcagno, Mauro. 2003. "Signifying Nothing: On the Aesthetics of Pure Voice in Early Venetian Opera." *The Journal of Musicology* 20 (4): 461–497. https://doi.org/10.1525/jm.2003.20.4.461

Cavarero, Adriana. 2005. *For More than One Voice: Toward a Philosophy of Vocal Expression*. Stanford: Stanford University Press.

Chadwick, Rachelle. 2021. "Theorizing Voice: Toward Working Otherwise with Voices." *Qualitative Research* 21 (1): 76–101. https://doi.org/10.1177/1468794120917533

Coole, Diana, and Samantha Frost (Eds.). 2010a. *New Materialisms: Ontology, Agency, and Politics*. Durham: Duke University Press.

Coole, Diana, and Samantha Frost. 2010b. "Introducing the New Materialisms." In *New Materialism: Ontology, Agency, and Politics*, edited by Coole, Diana, and Samantha Frost, 1–43. Durham: Duke University Press.

Connor, Steven. 2001. *Dumbstruck: A Cultural History of Ventriloquism*. Oxford: Oxford University Press.

Couzins, Richard. 2023. *Voice as Art: From Theatre to Forensics*. New York: Routledge.

DeLanda, Manuel. 2002. *Intensive Science and Virtual Philosophy*. New York: Bloomsbury Academic.

Deleuze, Gilles, and Felix Guattari. 1987. *A Thousand Plateaus: Capitalism and Schizophrenia*. Translated by Brian Massumi. Minneapolis: Minnesota University Press.

Dewey, John, and Arthur F. Bentley. 1949. *Knowing and the Known*. Boston: The Beacon Press.

Dolar, Mladen. 2006. *A Voice and Nothing More*. Cambridge, MA: MIT Press.

Dolphijn, Rick, and Iris van der Tuin. 2012. *New Materialism: Interviews & Cartographies*. Ann Arbor: Open Humanities Press.

Dunn, Leslie, and Nancy Jones (Eds.). 1994. *Embodied Voices: Representing Female Vocality in Western Culture*. Cambridge: Cambridge University Press.

Eidsheim, Nina Sun. 2015. *Sensing Sound*. Durham: Duke University Press.

Eidsheim, Nina Sun. 2019. *The Race of Sound: Listening, Timbre, and Vocality in African American Music*. Durham: Duke University Press.

Eidsheim, Nina Sun, and Katherine Meizel (Eds.). 2019. *The Oxford Handbook of Voice Studies*. Oxford: Oxford University Press.

Ernst, Wolfgang. 2012. *Digital Memory and the Archive*. Edited and with an Introduction by Jussi Parikka. Minneapolis: University of Minnesota Press.

Fast, Heidi. 2008. "An Emergent Tuning as a Molecular Organizational Mode." *Inflexions* (4): 342–357. https://www.inflexions.org/n4_An-Emergent-Tuning-as-an-Organizational-Molecular-Mode-by-Heidi-Fast.pdf

Fast, Heidi. 2016. "Singing Touches Shared Sensibility." Boldness Blog, Kone Foundation.
Fast, Heidi, and Milla Tiainen. 2018. "Voice." COST Action IS1307 New Materialism: Networking European Scholarship on "How Matter Comes to Matter." http://www.newmaterialism.eu/almanac/v/voice
Fischer, Clara. 2018. "Revisiting Feminist Matters in the Post-Linguistic Turn: John Dewey, New Materialisms, and Contemporary Feminist Thought." In *New Feminist Perspectives on Embodiment*, edited by Clara Fischer, and Luna Dolezal, 83–102. https://doi.org/10.1007/978-3-319-72353-2_5
Gamble, Christopher N., Joshua S. Hanan, and Thomas Nail. 2019. "What Is New Materialism?" *Angelaki: Journal of the Theoretical Humanities* 24 (6): 111–134.
Gatt, Caroline 2017. "Voices of the Pages." In *Knowing from the Inside: Anthropology, Art, Architecture and Design*, edited by Caroline Gatt. Aberdeen: University of Aberdeen.
Golovátina-Mora, Polina, Sunniva Skjøstad Hovde, and Tone Pernille Østern. 2022. "Seeding Actions." *RUUKKU – Studies in Artistic Research* 19 https://doi.org/10.22501/ruu.1395390. Accessed 30 August 2023. www.ruukku-journal.fi
Goodman, Steve. 2010. *Sonic Warfare: Sound, Affect, and the Ecology of Fear*. Cambridge, MA: The MIT Press.
Górska, Magdalena. 2016. *Breathing Matters: Feminist Intersectional Politics of Vulnerability*. Linköping Studies in Arts and Science no. 683. Linköping: Linköping University.
Gregg, Melissa, and Gregory J. Seigworth (Eds.). 2010. *The Affect Theory Reader*. Durham: Duke University Press.
Guernsey, Paul. 2017. "Post-Humanist Pragmatism." *Transactions of the Charles S Peirce Society* 53 (2): 246–269. https://doi.org/10.2979/trancharpeirsoc.53.2.04
Hammarström, Matz. 2010. "On the Concepts of Transaction and Intra-Action." In *The Third Nordic Pragmatism Conference*, 1–15. Uppsala.
Donna, Haraway. 1991. *Simians Cyborgs and Women: The Reinvention of Nature*. New York: Routledge.
Donna, Haraway. 2007. *When Species Meet*. Minneapolis: University of Minnesota Press.
Donna, Haraway. 2016. *Staying with the Trouble: Making Kin in the Chthulucene*. Durham: Duke University Press.
Harman, Graham. 2011. *Tool-Being: Heidegger and the Metaphysics of Objects*. New York: Open Court.
Haseman, B. 2006. "A Manifesto for Performative Research." *Media International Australia* 118 (1): 98–106. https://doi.org/10.1177/1329878X0611800113
Hemmings, Claire. 2011. *Why Stories Matter: The Political Grammar of Feminist Theory*. Durham: Duke University Press.
Hickey-Moody, Anna, and Tara Page (Eds.). 2015. *Arts, Pedagogy and Cultural Resistance: New Materialisms*. London: Roman & Littlefield.
Hird, Myra J. 2004. *Sex, Gender and Science*. Houndmills, Basingstoke: Palgrave Press.
Ingold, Tim (Ed.). 2022. *Knowing from the Inside: Cross-Disciplinary Experiments with Matters of Pedagogy*. London: Bloomsbury.
Irni, Kuura. 2013. "The Politics of Materiality: Affective Encounters in a Transdisciplinary Debate." *European Journal of Women's Studies* 20 (4): 347–360.
Irons, J. Yoon, and Grenville Hancox. 2021. *Singing (Arts for Health)*. Bingley: Emerald Publishing Limited.
Jenssen, Runa Hestad. 2022. "The Voice Lessons." *SCRIPTUM Creative Writing Research Journal* 9 (2). https://doi.org/10.17011/scriptum/2022/2/3
Kapadocha, Christina (Ed.). 2021a. *Somatic Voices in Performance Research and Beyond*. New York: Routledge.
Kapadocha, Christina. 2021b. "Introduction: Somatic Voice Studies." In *Somatic Voices in Performance Research and Beyond*, edited by Kapadocha, Christina, 1–11. New York: Routledge.

Kirby, Vicky. 2011. *Quantum Anthropologies: Life at Large*. Durham: Duke University Press.
Kittler, Friedrich. 1999. *Gramophone, Film, Typewriter*. Translated by Geoffrey WinthropYoung. Palo Alto: Stanford University Press.
Kontturi, Katve-Kaisa. 2009. "Eye, Agency and Bodily Becomings: Processing Breast Cancer in and through Images." In *The Future of Flesh: A Cultural Survey of the Body*, edited by Zoe Detsi-Diamanti, Katerina Kitsi-Mitakou, and Effie Yiannopoulou, 115–131. London: Palgrave Macmillan.
Kontturi, Katve-Kaisa. 2018. *Ways of Following: Art, Materiality, Collaboration*. London: Open Humanities Press.
Kontturi, Katve-Kaisa, and Milla Tiainen. 2024. "New Materialisms and (the Study of) Art: A Mapping of Co-Emergence." In *Genealogies and Methods of New Materialisms*, edited by Felicity Colman, and Iris van der Tuin. Edinburgh: Edinburgh University Press.
Kontturi, Katve-Kaisa, Taru Leppänen, Tara Mehrabi, and Milla Tiainen (Eds.). 2025. *New Materialism and Intersectionality: Making Middles Matter*. New York: Routledge.
Laasonen Belgrano, Elisabeth. 2018. "Ornamenting Vocality: An Intra-Active Methodology for Vocal Meaning-Making." *Ruukku – Studies in Artistic Research* (9). https://www.researchcatalogue.net/view/370801/370802
Laasonen Belgrano, Elisabeth 2019. "A Singing Orna/Mentor's Performance or Ir/rational Practice." *Ruukku – Studies in Artistic Research* 11. https://doi.org/10.22501/ruu.402061. www.ruukku-journal.fi
Laasonen Belgrano, Elisabeth. 2021. "Mapping the Burden of Vocality: French 17th Century Vocal Lamentations, Japanese Meditation and Somatic Intra-Action." In *Somatic Voices in Performance Research*, edited by Kapadocha C. Routledge Voice Studies: Research Monographs. Introducing chapter: https://www.youtube.com/watch?v=fjUtJ0_xgKs
Laasonen Belgrano, Elisabeth. 2025. "…Voicing as Loving as Living as Writing as Moving: Overflowing Knowledge Through a Sacred Rite of Passage…" (Editorial Prelude). In: *Special Issue: "Voice as movement and a spiritual 'in-between'". Journal of Dance, Movement & Spiritualities* 10(2). Intellect Publishing.
Laasonen Belgrano, Elisabeth, and Mark D. Price. 2025a. "Letting NOTHING Do Itself". In *Mattering Spiritualities*, edited by D. Mason and S. Battista. Routledge Press.
Laasonen Belgrano, Elisabeth, and Mark D. Price. (guest eds.). 2025b. Special Issu: "Voice as Movement and a Spiritual "In-Between"." *Journal of Dance, Movement & Spiritualties* 10(2). Intellect Publishing.
Latour, Bruno. 2007. *Reassembling the Social: An Introduction to Actor-Network-Theory*. Oxford: Oxford University Press.
Leppänen, Taru. 2018. "Always More than Two: Vibrations, the Foetus, and the Pregnant Person in Childbirth Singing Practices." *NORA – Nordic Journal of Feminist and Gender Research* 26 (2): 99–111. https://doi.org/10.1080/08038740.2018.1462255
Leppänen, Taru, and Milla Tiainen. 2018. "Trans-Becomings in Western 'Classical' Singing: An Intra-Active Approach." *Ruukku – Studies in Artistic Research* 9. https://www.researchcatalogue.net/view/372940/372941
Lorenz-Meyer, Dagmar, Pat Treusch, and Xin Liu (Eds.). 2019. *Feminist Technoecologies: Reimagining Matters of Care and Sustainability*. New York: Routledge.
Lykke, Nina. 2010. *Feminist Studies: A Guide to Intersectional Theory, Methodology and Writing*. New York: Routledge.
Macpherson, Ben. 2021. "The Somaesthetic in-between: Six Statements on Vocality, Listening and Embodiment." In *Somatic Voices in Performance Research and*

Beyond, edited by Christina Kapadocha, Kindle edition, 212–226. New York: Routledge.
Macpherson, Ben., and Konstantinos Thomaidis (Eds.). 2015. *Voice Studies: Critical Approaches to Process, Performance and Experience*. New York: Routledge.
Magnat, Virginie. 2019. *The Performative Power of Vocality*. Kindle edition. New York: Routledge.
Małecki, Wojciech, and Simon Schleusener. 2015. "'What Affects Are You Capable of?' On Deleuze and Somaesthetics." In *Deleuze and Pragmatism*, edited by Sean Bowden, Simone Bignall, and Paul Patton, 216–234. New York: Routledge.
Manning, Erin, and Brian Massumi. 2014. *Thought in the Act: Passages in the Ecology of Experience*. Minneapolis: University of Minnesota Press.
Massumi, Brian. 2013. "Prelude." In *Always More than One: Individuation's Dance*, written by Erin Manning, ix–xxiii. Durham: Duke University Press.
Mazzei, Lisa A. 2016. "Voice without a Subject." *Cultural Studies – Critical Methodologies* 16 (2): 151–161.
Mazzei, Lisa A., and Alecia Y. Jackson. 2017. "Voice in the Agentic Assemblage." *Educational Philosophy and Theory* 49 (11): 1090–1098.
Merten, Kai (Ed.) 2021. *Diffractive Reading. New Materialism, Theory, Critique*. Lanham: Rowman & Littlefield.
Neumark, Norie. 2010. "Introduction: The Paradox of Voice." In *VOICE: Vocal Aesthetics in Digital Arts and Media*, edited by Norie Neumark, Ross Gibson, and Theo van Leeuwen, xv–xxxiii. Cambridge, MA: The MIT Press.
Neumark, Norie. 2017. *Voicetracks: Attuning to Voice in Media and the Arts*. Cambridge, MA: The MIT Press.
Neumark, Norie, Ross Gibson, and Theo Van Leuuwen (Eds.). 2010. *VOICE: Vocal Aesthetics in Digital Arts and Media*. Cambridge, MA: The MIT Press.
Norbakken, Berit. 2024. "Grenseland - i stemmen," *Research Catalogue*. doct. diss. Tromsø: UiT Norges arktiske universitet - Norges Musikkhøgskole.
Norton, Kay. 2016. *Singing and Wellbeing: Ancient Wisdom, Modern Proof*. New York: Routledge.
Parikka, Jussi. 2010. *Insect Media: An Archaeology of Animals and Technology*. Minneapolis: University of Minnesota Press.
Parikka, Jussi. 2015. "Earth Forces: Contemporary Land Arts, Technology and New Materialist Aesthetics." *Cultural Studies Review* 21 (2). https://doi.org/10.5130/csr.v21i2.4317
Porges, Stephen W. 2011. *The Polyvagal Theory: Neurophysiological Foundations of Emotions, Attachment, Communication, and Self-Regulation*. 1st ed., Kindle. The Norton Series on Interpersonal Neurobiology. New York: W. W. Norton.
Rosa, Hartmut. 2019. *Resonance – A Sociology of Our Relationship to the World*. Cambridge: Polity Press.
Rosiek, Jerry. 2013. "Pragmatism and Post-Qualitative Futures." *International Journal of Qualitative Studies in Education* 26 (6): 692–705. https://doi.org/10.1080/09518398.2013.788758
Rosiek, Jerry. 2018. "Art, Agency, and Inquiry: Making Connections between New Materialism and Contemporary Pragmatism in Arts-Based Research." In *Arts-Based Research in Education: Foundations for Practice*, 2nd ed., edited by Melisa Cahnmann-Taylor, and Richard Siegesmund, 32–47. New York: Routledge.
Rowley, Glenn. 2020. "The Videos of Italians Singing Pop Songs on Balconies Are Fake." *Billboard* 16 March, 2020.
Seigworth, Gregory J., and Carolyn Pedwell (Eds.). 2023. *The Affect Theory Reader 2: Worldings, Tensions, Futures*. Durham: Duke University Press.
Shaviro, Steven. 2014. *The Universe of Things. On Speculative Realism*. Minneapolis: University of Minnesota Press.

Shusterman, Richard. 1999. "Somaesthetics: A Disciplinary Proposal." *Journal of Aesthetics and Art Criticism* 57 (3): 299–313.
Shusterman, Richard. 2008. *Body Consciousness: A Philosophy of Mindfulness and Somaesthetics*. Cambridge: Cambridge University Press.
Shusterman, Richard. 2012. *Thinking through the Body: Essays in Somaesthetics*. Cambridge: Cambridge University Press.
Sullivan, Nikki. 2011. "The Somatechnics of Perception and the Matter of the Non/human: A Critical Response to New Materialism." *European Journal of Women's Studies* 19 (3): 299–313.
Sullivan, Shanon. 2001. *Living across and through Skins: Transactional Bodies, Pragmatism, and Feminism*. Bloomington: Indiana University Press.
Tarvainen, Anne. 2018. "Democratizing Singing: Somaesthetic Reflections on Vocality, Deaf Voices, and Listening." *Pragmatism Today* 9 (1): 91–108.
Thomaidis, Konstantinos. 2019a. "Between Preservation and Renewal: Reconsidering Technology in Contemporary Pansori Training." *Theatre, Dance and Performance Training* 10 (3): 418–438. https://www.tandfonline.com/doi/full/10.1080/19443927.2019.1645040
Thomaidis, Konstantinos. 2019b. "Editorial: What Is New in Voice Training?" *Theatre, Dance and Performance Training* 10 (3): 295–302. https://doi.org/10.1080/19443927.2019.1677384
Tiainen, Milla. 2007. "Corporeal Voices, Sexual Differentiations: New Materialist Perspectives on Music, Singing, and Subjectivity." In *Sonic Interventions*, edited by Sylvia Mieszkowski, Joy Smith, and Marijke de Valck, 147–168. New York: Rodopi.
Tiainen, Milla. 2012. *Becoming-Singer: Cartographies of Singing, Music-Making and Opera*. Turku: University of Turku.
Tiainen, Milla. 2013. "Revisiting the Voice in Media and as Medium: New Materialist Propositions." *NECSUS. European Journal of Media Studies* 2 (2): 383–406. https://doi.org/10.5117/NECSUS2013.2.TIAI
Tiainen, Milla. 2017. "Sonic Technoecology: Voice and Non-Anthropocentric Survival in The Algae Opera." *Australian Feminist Studies* 32 (94): 359–376.
Tiainen, Milla. 2018. "Sonic Performance and Feminist Posthumanities: Democracy of Resonance and Machinic Sounds." In *A Feminist Companion to the Posthumanities*, edited by Cecilia Åsberg, and Rosi Braidotti, 103–115. New York: Springer.
Tiainen, Milla, Katve-Kaisa Kontturi, and Ilona Hongisto. 2015. "Preface: Movement, Aesthetics, Ontology." Edited Special Section "New Materialism." *Cultural Studies Review* 21 (2): 4–13.
Todd, Zoe. 2016. "An Indigenous Feminist's Take on the Ontological Turn: 'Ontology' Is Just Another Word for Colonialism." *Journal of Historical Sociology* 29 (1): 4–22.
Tondo, Lorenzo. 2020. "Coronavirus Italy: PM extends lockdown to entire country." *The Guardian*, 10 March 2020.
Tsing, Anna Lowenhaupt. 2015. *The Mushroom at the End of the World: On the Possibility of Life in Capitalist Ruins*. Princeton: Princeton University Press.
Tuana, Nancy. 2008. "Viscous Porosity: Witnessing Katrina." In *Material Feminisms*, edited by Stacy Alaimo and Susan Hekman, 188–213. Bloomington: Indiana University Press.
Uehara, Mayuko, and Elisabeth Laasonen Belgrano. 2020. "Performance Philosophy seen through Nishida's 'Acting Intuition'". In: *The Routledge Companion to Performance Philosophy*, edited by L. Cull and A. Lagaay (pp. 69–76). Routledge Press.
Van der Tuin, Iris, and Rick Dolphijn. 2010. "The Transversality of New Materialism." *Women: A Cultural Review* 21 (2): 153–171.
Wilmer, S.E., and Audrone Zukaskaitė (Eds.). 2023. *Life in the Posthuman Condition: Critical Responses to the Anthropocene*. Edinburgh: Edinburgh University Press.

Part I

Voicing ontologies

1 Voice

An imaginary figure of any thing

Elisabeth Laasonen Belgrano and Mark D. Price

Scene seven: Vocalizing heresies

Words pour out of my mind, heart, and soul,
Transforming into sound
And I know nothing else …

It happens on stage.
In life.
In prayer.

Sitting at the kitchen table. Trying to focus. Trying to collect thoughts that travel faster than the speed of light. They appear and they leave. Process had become Self. Has become Voice. A transient shadow of a physical body. Closing eyes. Recalling an initial purpose. Smiling. Intending to explain. Explaining without killing. It is the most challenging act.

There is always madness inside and alongside human vocalization, however *reasonable* a mask the speaker or singer may wear. A performative singer can play with all masks, and in that transformative playfulness, allowing them to confess their status *as* masks. The arts of this transformation pre-date recorded history, possibly emerging with the ability of our ancestors to mimic animals, but the processes also inform 'high culture'. For example, the semi-secret society called the *Incogniti* (Miato 1998; Calcagno 2003) may help us enrich our understanding of contemporary developments in new forms of materialist thinking. The *Incogniti*'s influence on the political and artistic world of 17th-century Venice was profound but also obscure. In keeping with their name, they are barely known among present-day philosophers and historians of ideas, much less among scientists or vocal artists. In this chapter, we aim to show how the ideas and practices of the *Incogniti* flashed up a series of strange resonances and enduring themes: not least of these are relevant to contemporary ideas of diffraction and voicing through time and space.

There is a tension in the gulf between historical forces and the sense we try to make of them. The application of fixed concepts is usually fatal to a process: "Everything that philosophers have handled, for thousands of years

DOI: 10.4324/9781003390992-3

now, has been conceptual mummies; nothing real escaped their hands alive [.] To them death, change, and age, like reproduction and growth, are objections–refutations, even" (Nietzsche 1997: 18).

Voice can call forth the presence of the past – not as mere nostalgia but as *live* performance and real material connection with spatial and temporal 'otherness'. A song bursts out of the logic of linear time and self-identity. It can disrupt the usual cognitive systems which operate in terms of either presence *or* absence. From within itself and at times seemingly effortlessly, the singing voice can provide a performative critique of a rational, closed, and anthropocentric self-hood. In what Nietzsche would call Dionysian intoxication, the permeable body of the singer is a transmitter, a way-station or vessel for the production of the im/possible: a sonic peacock's tail, inviting and provoking listeners to synesthesia: to likewise dissolve and enter into new entanglements and intra-actions.

With this in mind, the aim of this chapter is not a history of hard facts. Following the works of Morris Berman (1990) and Carlo Ginzberg (1992, 2004), we acknowledge that oblique, intuitive, and feminine strategies are required to explore a 'living' counter-history of ideas and arts. Many of these arts were from their outset aberrant and semi-clandestine. Ginzberg's book on 16th-century heresy opens with Celine's words: "Everything that is interesting happens in the shadows" (2004: epigram). How could this maxim fail to apply to a cabal calling itself the *Accademia degli Incogniti* – the Academy of the Unknown – working with a counter-philosophy of Nothingness and promoting public expressions of madness? What we know of their artistic and social aims suggests a project of fluidly somatic (which is to say, radically political) resistance to the fixed Logos. As with the contemporary Discordian anti-religion, any writings which claim to be definitive, true, or univocal about the *Incogniti* suggest a prejudice in favor of the dogmatic values the *Incogniti* opposed.

It is difficult to speak of heretics because almost all accounts come from their mortal enemies. A less obvious but more pervasive difficulty is that we inherit the languages and values and 'an entire grammar of thought' that has been conditioned by victorious orthodoxies. Whenever we speak coherently, we advance an unspoken majoritarian project. The unique, miraculous, refined, or rare experiences have to be 'made average' if they are to enter the economy of general concepts and ordinary communication. As William Burroughs said: "To speak is to lie, to live is to collaborate" (2014: 7). In Nietzsche's terms, languages are systems for expressing only the standardized affects, exchangeable and useful within 'the herd'. This is a problem when we approach any zone with a deep and complexly nuanced set of histories – such as materialisms, old or new.

Even so, some possibilities for vocalizing heresies remain open. We must think in a contorted manner and risk developing deformities: but we will speak of voices, whispers, and echoes in the side-lines of history which, despite the best efforts of orthodoxy, did not fade entirely into oblivion. We place these spirits of deformation in conversation with some key themes of 'the New Materialisms' – many of which are opposed to any talk of 'spirits' whatsoever. In going back to a mainly hidden and partly lost history of vocal art through the 'presence' of a singing mind and some oblique thinking, we hope to demonstrate how voices of a 'New Materialist' camp have allies and pre-echoes in older semi-materialisms: specifically, the European alchemy enshrined in early opera.

Scene six: Diabolus in musica

It makes me both worried and safe.
You ask from where.
From the sun I will tell.
From the monstrous spirits
Encountered in every corner of life
(smiling)
Double headed demons.
Opening their mouths
Wanting to eat every being.
Vomit them back into life.
Passing through.
Crossing over.
Voicing (im)possibilities.
Both-And-More.

How can one explain the 'wholeness' (or even holiness) of moving between notes?
Moving between worlds.
Moving step by step
without falling totally out of Self.

Karan Barad writes of "a performance of spacetime (re)configurings" offering "a way of thinking with and through dis/continuity – a dis/orienting experience of the dis/jointedness of time and space, entanglements of here and there, now and then, that is, a ghostly sense of dis/continuity" (Barad 2010: 240). Easy for her to say! But what does it mean? Are we to understand it as a Hegelian riff on the mutual implications of contraries and the logical problems of limits? Surely not, from a scientist. Or is Barad hinting that physical reality is poorly served by ordinary language, but can be corrected with improvised terms like (re)configurings and dis/continuity? There are reasons to suspect not for she opens a landmark essay with the words: "Language has been granted too much power" (Barad 2003: 801). Perhaps a useful tactic is to defer any strong demands concerning 'what a thinker means' and instead try to find out what the thought *does* when it is thrown into the alembic of lived experience and heated by performance.

Belgrano talks of the effects of a collision of fragments, a creative crash between early opera, her own vocal performance, and Barad's thought:

> Her words make me lose track of my own thoughts. Lose control. I am both continuing ahead of myself and at the same time, I become the very end of my own path. Becoming extremes and paradoxes in the same breath. Through the same voice. In the same scene. I am lost, yet woven into a net that makes me feel at ease somehow. I am still in my act of reading and performing, on the verge of madness. Or perhaps I am not mad at all. Like in the mad scene in Francesco Sacrati's opera from 1641, *La Finta Pazza* [...] In all this chaos, there is hope, keeping me on my feet. Barad continues to tell, 'what comes across in this dis/jointed movement is a felt sense of *difference*, of intra-activity' (2010: 240). Myself/my voice are part of yet another intra-active act, meaning here and now, including the Voicing you hold in your hand, your eyes reading these words, your thoughts meeting mine on the page in front of you.
>
> (Belgrano 2016: 185)

To which the skeptical reader might respond – all very well, but that does not help us to *understand* what is going on here! – and that would not be an illegitimate response. We will try to address it by highlighting a connection between philosophy as a *practice* of thinking/writing/reading and vocal performance.

If we consider philosophy as a special subset of performance, the question of understanding is commutable. In her ground-breaking volume of 2008 on "the transformative power of performance", Erika Fisher-Lichte asks – "Can performance be understood?" She answers via a meticulous investigation of the strategies used in live performances and 'arts events', but also in everyday living. What emerges in her analysis is a specific account of the auto-poetic feedback loops of producing and receiving meaning and knowledge: and these are based on the performative sharing of movements and actions between actors and audience. Fisher-Lichte writes:

> Performance, then, requires two groups of people, one acting and the other one observing, to gather at the same time and place. [The actors] move through space, gesture, change their expression, manipulate objects, speak, sing. The spectators perceive their actions and respond to them. Although some of these actions might be limited to internal processes their perceptible responses are equally significant: the spectators laugh, cheer, sigh, groan, cry, scuff their feet [...] eat and drink, crumple wrapping paper, whisper or shout comments [...] in short, whatever the actors do elicits a response from the spectators, which impacts on the entire performance. In this sense, performances

are generated and determined by a self-referential and ever-changing feedback loop. Hence, performance remains unpredictable and spontaneous to a certain degree.

(2008: 38)

Beyond that certain degree of unpredictability, there is yet more. Two human groups of actors and audience are involved, but these are inter-meshed and cross-cut with other interactions and intra-actions from non-human forces. Some of these we have names for, and are obvious to the humans in a performance environment – temperature, humidity, the hour of the day or night, and so forth. We may assume there are many others without names and not so easily available to sense.

If we speak of the *Incogniti* and try to vocalize their heresy, we must do so by oblique strategies and sly means. Even their heresies were multivalent or paradoxical: they were, in many cases, *both* heretics *and* confirmed Catholics. There is much we have to guess at, using 'history' and 'facts' to show something present and vital, even if it can only be seen with sideways glances or heard with slanted ears.

The *Accademia degli Incogniti* was founded in 1630 by Giovani Francesco Loredan and Guido Casoni. Its role in artistic and political life was as large as it was shady. As best we can discern, the philosophical impetus came from a mixture of neo-Aristotelianism and the re-invigorated Hermetic tradition channeled via Marsilio Ficino. They often used ciphers and coded symbolism and tended to publish anonymously. Most were men, and most of high social status, with enough economic muscle to build their own theater. They directed not only individual operas but a large part of the cultural climate of Venice (Rosand 1991).

Rather than assuming their secrecy was entirely sinister and manipulative, we should recall that Papal authority – though weaker in Venice than many cities – was a changeable force. Free-thinking philosophy and progressive arts were risky endeavors. Early opera presented pagan divinities, illicit sexual relations, sorcery, spirits of doubtful sanctity, and gender swapping: Sacrati's *La Finta Pazza* (1641) had most of those elements. Monteverdi's *Orfeo* (1607), sometimes described as the root of all opera, involves communication with the dead.

Popularizing this new and hybrid artform was a high-stakes gamble against religious powers which were unlikely to be mellow in their response. Had the temper of the age been slightly different, the ringleaders would have been excommunicated or worse. As the dice fell, their works found powerful supporters. The public loved it: many were moved to tears. The commercial classes saw the opportunities and invested in purpose-built opera houses. The Church was bi-polar about the phenomena. On the one side, condemnations promising hellfire for all involved. On the other side, Cardinal Mazarin was pivotal in organizing the performance of *La Finta Pazza* in Paris in 1645. This is highly peculiar because the mass of testimony concerning Anna Renzi's performance in *La Finta Pazza* describes nothing less than *the material transformation of the performer's body into a conduit for non-human forces.*

Scene five: Intermezzo

> *Born in the very act ...*
> *Intra-action.*
> *Diffracting patterns ...*
>
> *What is the purpose*
> *For trying to define*
> *Anything ...?*

Singing was feared by the Church, and the roots of that fear ran deep. People could all too easily lose themselves in song. They could lose their souls. Song disrespects boundaries and connects seemingly unconnected bodies. As early as the 3rd century AD, Saint Cyprian decried the use of song in worship, writing that voices, if heard at all, should be modest, quiet, and clear. In the words of Monseigneur Charles Pope (2017):

> There was in the early Church a very persistent theme that music itself was problematic. Many ancient bishops and Fathers of the Church barely tolerated it, sought to limit its influence, and/or were deeply suspicious of any singing at all. In his essay "On the Theological Basis of Church Music," Cardinal Ratzinger (drawing from sources such as Pope Gregory the Great, St. Jerome, Gratian, and even as recent as St. Thomas Aquinas) describes the rather negative opinion in the early Church of any music involving instruments, harmony, or anything deemed "theatrical." He writes, "Instrumental music, understood as a Judaizing element, simply disappeared from the early liturgy without any discussion; the instrumental music of the Jewish temple is dismissed as a mere concession to the hardness of heart and sensuality of the people at that time".[1]

How the Jews managed to be both hard-hearted and sensual is an anti-Semitic mystery, but they figured as necessary specialists in Venetian mysticism and metallurgy, to which we will soon return. For now, we merely note that: "Venetian rabbis such as Simone Luzzatto, Salomo Hai de Seraval and Isaiah Hayyim ben Eliezer were solicited by the *Incogniti* for learning and debate and were probably present at Academy meetings" (Chayes 2017: 75). The consensus of Saints and Popes held that if religion *must* have music, then it should exclude plurality of voices and avoid ambiguity. It should be hostile to multiple interpretations, for nuance breeds heresy. It should evoke no emotional or somatic content except cool devotion.

But music, and especially song, dissolves any assumptions of hard barriers between separate bodies. Intractable and ever-popular, the music of the human voice creates somatic intimacy and in this sense it is sacred. The understandable fear of the Church was that song does this without any regulation from the 'proper authorities'. It is axiomatic for most monotheistic religions to seek a monopoly on contact with the sacred. One God, one true religion, one body per life, and one soul per body: all else is of the Devil.

Yet the *sacred* differs from the religious. As that basest of materialists Georges Bataille noted, the sacred is opposed to religion because "intimacy cannot be expressed discursively" (1990: 50). Orderly religion can give people the *work* of salvation and the calm of clear consciousness, but what is 'intimate' in Bataille's strong sense "is what has the passion of an absence of individuality" (Bataille 1990:50). "The sacred [...] calls for the general negation of individuals as such" (Bataille 1990: 53). To be transported by music is to feel a sacred union and/or emptying out and/or multiplication of the ego. It is mortally dangerous to the economy of orthodox spirituality. If your spiritual children are 'called beyond themselves' and 'lost in music', what entities might creep into their bodies when their souls are out dancing?

A transported singer may become a monster with one soul in several bodies, or with multiple 'selves' in one body (Belgrano 2011). And in worse/best case scenarios, this spreads to the audience. This is a more or less precise description of the contemporary practice of many vocal artists – perhaps most famously Tom Waits and Diamanda Galas. And as we shall see, it was a stated mode of procedure for Anna Renzi and the *Incogniti*.

Scene four: When metals flow

If I try to fool you
Why would I ever do such an evil?
I am telling the truth.
And my voice is the voice of a messenger.
Delivering holy water.
Sacred water.
Living water.
Tears.
Laughter.

The Loredan/*Incogniti* crest resembles an open mouth (Calcagno 2003; Belgrano 2011). At its center are the multiple mouths of a river, possibly the Nile flowing out of Egypt. Our word alchemy comes from the Greek *khemeioa* or *Khemia*, the Greek name for Egypt, and *khymatos* – that which is poured out. Populist figurations of alchemy tend toward its dismissal as a proto-scientific folly or else refigure it as a psycho-spiritual quest of the Jungian variety. Both positions eclipse key aspects of alchemy as a pan-European force for political change as much as personal or chemical transmutation. Gender politics was at the fore:

> In 1581, the Venetian writer Moderata Fonte, borrowing the imagery of metallurgy, wrote eloquently of women's untapped potential in the

> liberal arts as "buried gold", needing only to be unearthed and given the same attention as that afforded to men: "Gold which stays hidden in the mines/is no less gold, though buried,/and when it is drawn out and worked/it is as rich and beautiful as other gold."
>
> (Ray and Lowe 2015, introduction)

The immense political potentials and dangers involved in alchemy remain largely overlooked.

The word 'dabbled' is trivializing and contemptuous, so it is vexatious to read the oft-repeated absurdity that Monteverdi 'dabbled' in alchemy. As with music and the Jews, alchemy occupied a liminal place in Renaissance Italy. Experimental studies were partly encouraged and often repressed, but always a dangerous pursuit. The alchemist Marco Bragadino was at first welcomed by Venice, then expelled and executed in a horribly botched manner in Padua in 1591. Cristoforo Parigino was fined heavily and exiled. Like most of the *Incogniti*, Monteverdi was socially well-connected. He was also an ordained deacon of the Church. Money and status *sometimes* deflected the full force of the city authorities ('The Forty') and the Inquisition. Sometimes it did not, and Venetian law could be horrifically theatrical in its "elaborate ritual executions that included public mutilation" (Ruggiero 1978). Even the best-connected of the *Incogniti* would no more 'dabble' in alchemy than today's Saudi princes or wealthy Chinese merchants would 'dabble' in the manufacture of LSD. We can infer that Monteverdi and the alchemically inclined *Incogniti* had high levels of commitment.

The analogy with LSD and psychedelics is apt. Much as the bohemian artists and intellectuals of the 1960s sought to 'turn on' and transform society with extraordinary cultural-chemical experiences, so too the work of alchemy promised huge changes in the consciousness of its practitioners and audiences in tandem with changes in the material elements in the alembic:

> Certainly opera could on occasion draw the audience into a kind of mystic participation, not resembling the Mass, so much as the sacred drama enacted at the Eleusinian Mysteries. After an immersal of three or four hours in an altered state of consciousness, some might find themselves indelibly marked by the experience, while none would easily forget it.
>
> (Godwin 1990)

There is no doubt that Monteverdi and his associates utilized alchemical ideas to develop modes of vocal expression, which were shockingly agitating to the listeners of his day. Written accounts by the earliest listeners of Anna Renzi's performances of Monteverdi's works testify that they had been profoundly changed. A volume of poetry was published in honor of her performances (Strozzi 1644). Giulio Strozzi declared Renzi's performances 'divine' and was moved to write:

> *Hearing your mournful voice,*
> *Your sweet words, your endearing expressions,*
> *Just as they filled our breasts*
> *With pity, ah, well do I know that*
> *Nero would have been rendered humble*
> *and compassionate.*

Monteverdi's letters of the 1530s claim that he was following nature: but he is working from a very different model of nature than the mainstream Christian one. The Papacy could make use of Robert Fludd's ideas of a *Musica Mundi* with its monophonic deity and orderly choirs of angels 'singing with one voice'. But Monteverdi's use of discord expanded the possibilities of expression and reconfigured what song was, what it could do, and what singers could be. For example, his *Duo Seraphim* is a disharmonious invocation in which one can almost hear the Jews and Moors of King Alphonso's Spain contesting with Catholicism. It is utterly pious but almost aggressively material in its vocality, with much repetition of seemingly nonsensical phonemes and breathing patterns to strain the lungs of all but the most adept. Waves of painful, combative yearning build toward the crowning 'Sanctus!', pushing the human material of his performers and listeners into entirely new territories. We might also think of the final duet in his *Poppea*, in which the voices of Poppea and Nero transform from air to sound, then melt separate sonic units and personalities together until dissonance becomes resonance. Monteverdi's music rattled the 'great chain of being' so violently that as early as 1600, cultural power-brokers such as Giovanni Artusi devoted time and money to condemning him in print (Artusi 1600, Cusick 1993).

The *Incogniti's* commitment to creative discord and their not-quite-monotheistic model of Nature capable of 'immanent' activity might be viewed as a close relation to ideas within the New Materialisms. More interestingly, they resonate with several centrally stated 'aims' of New Materialist thinking. New Materialism aims at transformation: "calling upon us to reorient ourselves profoundly in relation to the world, to one another, and to ourselves" (Coole and Frost 2010: 6). It aims to focus on the "active processes of materialization of which embodied humans are an integral part" (Coole and Frost 2010: 8). There is a strong de-emphasis on ontological opposition in favor of a stratified differential ontology with "no definitive break between sentient and non-sentient entities" (Coole and Frost 2010: 1). All of which would be agreed upon by the overwhelming majority of alchemists. None of this makes alchemy 'the same' trajectory of thought as our New Materialisms, but the similarities are more than superficial. The jewel in the crown of these similitudes is the aim to "depart from the prioritization of the subject" to break through the limits of the human (Dolphijn and Van der Tuin 2012: 162).

It is impossible to determine whether the above points indicate the return of the repressed, the continuance of a perennial undercurrent, a Baradian (dis)continuity, a 'mere coincidence', a case of retro-causality, or a magical bunny-hop in history. We present the dots: join them and make constellations as you will.

Scene three: Nothingness and pure voice

Cantatrice beltà mentre diletti
Desti ne'petti altrui fiamma vorace
E tormentando il cor, l'udito alletti.

Beautiful singer while you are delighting
you awaken a voracious flame in the hearts of others and
tormenting the heart, you entice your listeners.
(BENEDETTO FERRARI DELLA TROMBA, in Strozzi 1644: 28)

Mercury is a lively element, having no shape but becoming any shape in accordance with its environment. It has no color but reflects all colors around it. In this sense, it resembles consciousness, and resemblance underpinned much of the alchemical world-view with its vision of 'essential signatures'. The resemblances between mercury – consciousness – Nothingness – and voice are the keys to a Pandora's box of mad and monstrous but also joyful and ecstatic ideas. These ideas cannot be separated from vocal practices which fit the same description, with transformative 'madness' at the front and center stages.

The efforts of the orthodox to reduce the voices of the *Incogniti* to nothing were almost comical, like a cop delivering a hard kick against an open door. For these heretics had already unlocked the portal of Nothingness. Pre-figuring themes in 20th-century existentialism, the *Incogniti* made 'Nothing' their home territory. Their very name suggests a relationship to non-knowledge, a voyage into sacred unknowing. Their use of Nothing was not merely tactical; it was a foundation or an anti-foundation in the abyss, a non-position. This primacy of No-thing-ness has mythical analogues in the trickster gods Set and Loki, the night-shaded goddesses Hathor and Hecate: interesting misfits who could not or would not be constellated into the heavens. Also the Native American Coyote, the howling and laughing shape-shifter. But most of all Tawusi Melek, the Peacock Angel of the Abyss from the Yezidi religion. An emanation of God descended into matter, Tawusi Melek is the archangel of light and darkness and all colors in between. The Yezidis were deemed Satanic by other monotheisms and violently persecuted from 630 AD to the present (Allison 2017). The creative fecundity of both matter and Nothingness tends to catalyze extreme reactions.

Let us sketch how it was possible for the *Incogniti to* figure Nothing as a pure potential linked to pure voice. Nothing, like alchemical mercury, could become 'anything'. From this seed, a million heresies bloom and riot.

One obvious theological implication is that the unlimited creative potential of Nothing is divine. The 'supreme creator' is the supreme Nothingness. If Nothingness becomes a (non-)symbol for God, we open doors to speculations which weave together various strands of atheism, polytheism, and/or theism. The relationship proposed by the *Incogniti* between vocality and Nothingness can be simplified to the fact that words were the sparkles of meaning, and when spoken, the rational context is transformed into other meanings and purposes. Language seems to work of its own accord, deflecting and diffracting even the strictest monotheism and its wager on 'literal scriptural meaning'.

In 1634, the *Incogniti* published a discourse on Nothingness, *Le Glorie del Niente* (Manzini 1634). They were playing, very *seriously playing*, with negation. The logical moves might look familiar to readers of Hegel, Heidegger, and Derrida, though the *Incogniti* lacked the same high-resolution piety and humorlessness. For the *Incogniti* tricksters, Nothingness was, in a sense, fragmented, but at the same time, it was a timeless or primal unity filled with paradoxes and multiple meanings. A list of figures of Nothingness was formulated. Voice was one of them.

Their joke was an insistent one: *Nothing matters*. Nothing(ness) gives; it is active and creative. It seems a short leap from *Nothing matters* to the extended New Materialist interrogation of 'How Matter Comes to Matter' – the title of an essay by Barad and the theme of an international conference on sonic arts and New Materialist ideas in which Rick Dolphijn and Iris van der Tuin explored the relation between human and non-human lives, via the arch-heretics Spinoza and Nietzsche (Dolphijn and Van der Tuin 2012). And what of voice as a symbol of nothing and as a thing which matters immensely? Voice is the mercurial, hermetic go-between for interior and exterior, human and environment, being and becoming. Currents of subtle material running in and out of our faces suggest strange realities literally going on unnoticed 'beneath our very noses'.

Unlike the transcendent *Logos*, voice is inseparable from active matter – and the matter may or may not be 'living' or organized into a human body. Matter has agency, 'is' agency, not the passive clay or chaotic water of Biblical mythology awaiting instruction from above. As Barad has it:

> Agency is not aligned with human intentionality or subjectivity. Nor does it merely entail resignification or other specific kinds of moves within a social geometry of antihumanism. Agency is a matter of intra-acting; it is an enactment, not something that someone or something has. Agency cannot be designated as an attribute of 'subjects' or 'objects' (as they do not pre-exist as such). Agency is not an attribute whatsoever – it is 'doing'/'being' in its intra-activity [i.e. a performativity]. Agency is the enactment of iterative changes to particular practices through the dynamics of intra-activity.
>
> (Barad 2003: 826–827)

The voice in song also differs from the 'Logos' in its radical fluidity, imminent plurality, and sensuality. Song is not only resistant to 'being' but actively aggresses against it. The religious dogmatists were right to fear it as a portal for 'other forces'. Even without demons, the singer's body itself is seen as a contaminant, with ever-present sexual overtones. Pier Francesco Tosi urged singers to aim for 'noble simplicity' in order to "preserve music in its chastity" (Bethell 2021). Thus, it is ironic that the often wordless, affect-rich ornaments and ululations of the operatic voice became known to musicologists as 'pure voice' (Calcagno 2003).

Pure voice is anything but pure. It teems with hybrid forces, bastard energies, glitches, seeds, and viruses. Given the right conditions, these grow and mutate without the intervention of the singer's consciousness, for pure voice offers portals to worlds outside the human ego as it diffracts into Nothingness (Laasonen Belgrano 2020). To draw on Bataille's distinctions between the religious and the sacred, pure voice abrades religiosity not because it is secular but because it actually delivers the promise of restoring the intimacy of the sacred. The bursting lament overspills and calls forth tears. The passionate sigh sexually excites the listener. It does so in 'this' world; it expands the world in ways one can actually experience – not as a promissory note against good behavior to be redeemed in some post-mortem realm.

The emergence of the operatic voice in Renaissance Italy involved a complex set of cybernetic loops between somatic practices, philosophy, theology, popular culture, commercial and political interest, mass psychology, myth, and the supposedly 'internal' realms of sensations, passions, and aesthetic responses. And it is clear that the machinery and the monstrosity of this emergence were 'not under central control'. It was diffracting, mutating, performing in ways unforeseeable to its creators – not all of whom were human, anyhow.

In speaking of the 'operatic' voice, we aim to gesture toward the art of specific composers and performers. But we also wish to expand the concept beyond the realm of 'operatic vocalizing' in its most common sense, which is so culturally over-determined that it is easily parodied as bombastic warbling. Consider 'operatic' not as a description of a genre but in closest accord with the etymology of the word 'opera'. An operatic voice – 'the voice of highly operative forces'. A voice which works real somatic, affective, and material changes in the body of the performer and listener. Laughter and tears are but two of its limitless effects, and even those two options are massively multivalent.

There are analogues in the Brahmanic *jalal* voice that emerges from Mauna, a silence which has a voice of its own (Rig Veda VII, 104.1, 10, 11). In Taoist philosophy, 'Nothingness' is not inert, but the completely elusive absence which is the precondition for there to 'be' anything whatsoever. We make no claims on those exotic figurations as influences on the *Incogniti* any more than the words of recent popular songs: "Listen to the silence. Let it ring on [...] No language, just sound, that's all we need know" (Curtis 1979).

We merely note resonances, diffractions, and the ubiquity of a seemingly nonsensical or paradoxical zone with huge affective reach. Artists, thinkers, and writers of the most varied times and cultures recognize the fecundity of Nothingness and the silence which speaks volumes. To stop there would be too humanistic. Matter voices silence too: and while that point may be drawn to our attention by physicist-philosophers such as Barad and poets like Whitman (Marsden 2020), few people would consider the limits of human attention to be identical to the limits of the world.

All very well, but what do such ideas 'do'? What can they change? How do they function? Such matters are best handled by a vocal performer, who speaks of acts which are her own yet also violently transpersonal and trans-human:

> The questions pile up in the singer's *body* and thoughts. Methodologically, her voice leads her through the questions, and she experiences answers that *are not always explained*. Answers that can work oracles but which constantly open doors to new insights, insights that guide her in shaping a maze of new realities [.] on a stage, strategically placed, on neutral ground, as if in a vacuum. The lament together with the act of insanity in the first operas of the 17th century was lifted to the skies by the audience. From a distance, through actions [the distance was collapsed into intimacy] the unthinkable and the unrealistic became clearly intrusive and real [...] She literally jumped between realities, physical, mental, and vocal [...] she got her audience to get lost in the real.
>
> (Belgrano and Price 2019– ongoing, personal communication)

Scene two: Spirits and mutants, monsters and madness

It all comes through seed.
Tiny seeds.
Slowing becoming fruits of imagined contemplations.
I would never know in advance.
They happen.
Connecting through the movements
of the stars in the sky.
All elements contribute to the sound of angels.

They call me my many names.
Siren.
Thalia.
Melpomene.
The Fourth Grace.
Lover.
Venus.
Soldier.

Helena.
A fool.
A monster.
A man.
A woman.
The monstrous Signora Anna Renzi romana.

Before the theoretical elaborations of 'dead matter' came to dominate the sciences, it was widely accepted that metals grew in the earth as much as plants grew. Metals could be 'seeded' and the task of the alchemist was like that of the agriculturist, speeding things toward their full potential. They had few doubts about "the world's radical aliveness" (Barad 2007: 33). Alchemy was a very broad non-Church, but most practitioners were concerned with the materialization of spirit and the spiritualization of matter. Fasting, chanting, and other modes of consciousness alteration were as much a part of the work as the crucible and the forge. Where the operatic alchemists 'went wrong' in terms of New Materialist evaluations is their (mis)recognition of the mirage of 'spirit'.

But 'spirit' can also be articulated in materialist-monist registers: Nick Land's essay "Teeth and Spirit" (2011: 65–73) develops precisely this thought of spirit in terms of werewolves, rats, and plagues. For Land, these products of contaminated blood represent a volatilization of matter which takes us *beyond the human* in all directions. They crawl under or run in circles around any presumed evolutionary hierarchy. In Land's work, 'spirit' is a vector of uncleanliness and ungodliness, a resistance movement to transcendence. Alphonso Lingis contends that the primal Gods were animal spirits, later semi-humanized, and the 'mute idols' of material spirits remain active in contemporary capitalism (Lingis 2000). Nietzsche's eminently materialistic elaboration of tragedy (1872) emerges from the 'spirit' of music.

Thus, we do no great service to philosophy by trying to police materialism, new or old, by tolerating the word 'spirit' only if it is understood as a metaphor. Such a move is possible only if we deem other aspects of language to be non-metaphorical: as if terms like 'matter', 'gravity', and 'electromagnetism' were identical with reality rather than the useful chatter of technically equipped primates. Beyond our words and theories, there is wild matter.

> That matter is volatilised to different degrees of spiritualisation is not in the least dependent upon spiritual causalities of any kind. Between the wilderness and the polis is a wilderness history - a genealogy - and not a political history. Regression is not an undoing of the city's work, but a recurrence of *impersonal creativity*.
>
> (Land 2011: 69)

The uncanniness and disturbing delights of this 'impersonal creativity' emerge as materialized spirit/volatilized matter in all its impurity via 'pure voice'. By chance – or perhaps not – some of the best ways of describing 'pure voice' can be lifted directly from New Materialist thinkers who were not directly speaking of voice at all, but rather of their methodologies. It is "endlessly open [...] to a variety of possible and impossible reconfigurings" (Hinton 2013: 182). It is "blasting, bursting open, and scattering [...] to effect a complete reorganization of meaning" (Barad 2007: 41). It is what happens in a live environment when matter, Nothingness, singer, and audience bloom into a fuller and 'even more alive' environment: an ecology in which worlds and meanings can create themselves. We could ask what it means but to paraphrase Land – the important thing is 'epidemics', not hermeneutics.

The mad scenes of early opera invoked an epidemic of affects, meanings, mythologies, and characters until the overflow became too much for the human: Anna Renzi specialized in vocalizing the fullest range of clean and unclean spirits in her performances. In our ultra-capitalized and cynical era, it is easy to jump to the thought that the *Incogniti* were trying to make money from show-business. They already had money, enough to build opera houses and support droves of artists. It seems more plausible that the aim of the *Incogniti* was to spread the multiplication of spiritualized bodies and material spirits into the populace at large.

Signora Anna Renzi's voice has become a myth that can function analogously to Nietzsche's refiguration of Dionysus, or the liminal semi-pagan Holy Grail. Its scripting by Strozzi and Monteverdi was long in preparation, but the preparation was geared toward madness and unpredictability. It was commonly held that man has one nature and one person. One soul per body, except: "In any given pregnant woman however there are more persons in more natures, and vice versa" (Marsilio Ficino 1474). In Renzi's performances, the singer's voice functioned as a feminine matrix in a state of permanent and multiple pregnancies. It birthed and burst and blossomed with powers and personalities.

After a century of psycho-analytic variations, it is hard not to read Freud backward into Renzi's remarkable facility for ego-swapping between man, woman, god, goddess, madwoman, criminal, victim: easy to reduce the phenomenon to a mental illness made functional by sublimation into art. Such recourse to a 'tameable' unconscious is one way of making sense of what does not make sense, or overflows as 'too much sense'. But the reductive thought seems flat and barren in comparison to Land's suggestion that "If the unconscious is structured like a language it is only because language has the pattern of a plague" (Land 2011: 68). When she sang, Anna Renzi was not herself. Contagion ensued: she and the audience were transformed. This transformation is a matter that extends far beyond theory, concepts, or anything ending in – 'ism'. Articulating the effects and forces in a materialist (or any other) register can never adequately extend to the horizons of a 'lived encounter' with madness and the other-than-human. Language deals only with the most common perceptions.

Apart from romantic love, alchemy was one of the few 'other spaces' where willfully violent imbalances were deemed no insult to either Nature or God. Alchemical transformation begins with deliberate sickening and corruption. This initial *nigredo* or blackening phase involves not only the degradation of matter, the *primum materium* – it also requires a sympathetic degradation of the alchemist's psyche. Images for this deliberate erosion of the self include kings being clubbed to death or devoured by wild animals, torture, mutilation, the sun turning black or drowning. Breakdown was fervently desired. Madness, and especially what we would now call chronic depression, was the price of the ticket to enter. Once inside the alchemical venue, all hell and heaven might break loose (Marlan 2008).

We face a problem too large to deal with in detail here but cannot go unremarked. In the world of the *Incogniti*, there were aberrant mental states, but they were not yet annexed by the medical sciences: madness 'was not madness' in today's sense. Renaissance Italy had no clearly analogous concept to 'mental illness'. Imbalanced humors or *strano* are perhaps closest to what we would now consider a 'diagnosis', but a variety of terms were in circulation: *frenetici* (frantic), *deleri* (delerious), *pazzia* (madness, related to 'impassioned'), and *follia* (folly or error). Whatever the phenomena were and whatever names they went by, there was a buzz of deep strangeness around the new art of opera. Masses of ordinary people wanted to see it, hear it, and live a while in the un-sane zone (Cama 2014).

> As with so many other operatic conventions, we owe the critical as well as musical articulation of the issue of madness to Monteverdi [who] recorded his ideas on the subject in a famous series of letters of 1627 concerning an opera on which he was working with Giulio Strozzi, *La Finta Pazza Licori* [...] In advising Strozzi about the kind of text he wanted, Monteverdi was particularly concerned with the character of Licori, the 'finta pazza' herself, and especially with her madness.
>
> (Rosand 1991: 347)

Monteverdi envisioned the central character of *La Finta Pazza Licori* as self-multiplying, like alchemical gold, transforming and transformative: paradoxically one and many, a unity and a plurality, every body and no body. And though *La Finta Pazza Licori* did not gestate to the theater, it directly prompted Sacrati and Strozzi's *La Finta Pazza* of 1641. In figuring the singer as a metamorphic substance and a vector of non-human (more-than-human) forces, the exponents of early opera engaged in politics in the widest sense: a politics of the body which challenged the social and theological orthodoxies of Europe. Perhaps the most striking zone of contention is that of sex and gender in relation to the Divine Order.

Monotheism generally places heavy emphasis on the sanctity and immutability of natural kinds and widely holds that the most basic distinction after dead or alive is 'bodily sex'. The ubiquity of the image of the hermaphrodite in alchemy evidences a different understanding of living bodies (Roob 2014: 370–377). The hermaphrodite and/or the androgyne – often monstrously two-headed – displays the possibilities and the desirability of bodies flowing into each other. For those invested in the ideas of fixed categories, values, and beings, the hermaphrodite appears as a carnal monstrosity, a mistake, a crime against God and Nature. From an alchemical perspective, the breaking down and merging of 'natural kinds' was the path to perfection: "First we bring together, then we putrefy. We break down what has been putrified. We make pure what has been divided. We unite what is made pure and harden it. In this way is One made from man and woman" (Mozart et al. 1790). 'Alchemical opera' extended far beyond its cradle in Monteverdi's Italy. The narrative core of *The Magic Flute* "is an enactment of the alchemical *opus magnum*, in the form of a *chemical wedding*. Towards the end of the 18th century, alchemy was still a prominent mystical current within the Order of Freemasons of which Mozart and his librettists were members" (van den Berk 2004: 287).

Scene one: Voice as material operator/disruptor

A voice. A body. A human.
A God/ess carries many names.
Only fools try to make anything reasonable about me.
It can't be done.
Because No *Body knows what will come.*
Not even Self.
Not until it happens.
Now. Now. Now.
Again. Again. And again.

This is
A confession.
A meditation.
A contemplation.
An act of transformation.
An imagination.
A time-space-traveling.
A performance.
A magical spell.
A performance.
Again.
A wonder.

It would be only slightly monstrous to retro-fit Terence McKenna's ideas of techno-shamanism or Donna Haraway's cyborg feminism into the realm of

Venetian opera. The generative roles of the feminine and the machine were certainly present, front and center stages:

> For musical and mythological reasons, women play as important a part as men in these dramas, wielding an influence that they seldom enjoyed in society. Sexual ambivalence is rife in the multiple travesties, disguises, and in the regular casting of men (the Castrati) in female roles, and women in male ones. And above all, there was the spectacular element, which made the designer of scenery and machines often the highest-paid contributor to the enterprise. It was his illusionary skill that, like the conjuration of a magus, summoned the archetypal beings and their worlds into sensible existence.[2]

New materialism, old materialism, and even older materialism: the shock of the new, the shock of the old. Was there ever a time when a percentage of people were 'not' trying to summon "archetypal beings and their worlds into sensible existence" (van den Berk 2004) via machinery and sexual transgression? A grinding bowl and a stick, paint and a cave wall depicting animal/human hybrids, a shaman cross-dressed between genders and species, deer or bear, beating a skin drum: these too are 'technologies' courting libidinal strangeness, *strano*, and real affective connections. From scant evidence, we might imagine the metaphysics of ancient tribal peoples and speculate that they subscribed to some kind of ontologically divided 'spirit world' and a material one. We might even enjoy a tingle of superiority that our presuppositions and technological operations 'prove' how wrong they were. But the 'high-culture' works of early opera are continuous with allegedly 'primitive' practices, as are the 21st-century projects aiming at thoroughly materialistic artificial intelligence. We are addicted to summoning alien powers and non-human beings into existence. Monotheistic religions have tried hard to stop this sorcery. So too reductive materialism. It would take more than one or two gods to stop us.

In various contexts, the voice has been considered as a technology for the creation and/or conjuration of non-human life. The Golem of Prague was dead clay animated by written language, but the creatures of the Italian Hermeticists were figured as *homunculii* or *daemons* or artificial souls woven by the motions of the element of air. As Ficino writes, the material of harmony "is air, hot or warm, breathing and somehow living, composed like an animal of certain parts and limbs of its own, like a living thing, not only possessing motion and displaying passion but even carrying meaning like a mind, so that [song] can be said to be a kind of aerial and rational animal" (1489: chapter 21).

This sonic re-organization of matter differs from the orthodox theological view of the *Logos* or a lone Divine Craftsman pressing form into something inert. In some Hermetic accounts, we find mobile, active matter (the blood) working in concert with a subtle substance (spirit, which is not identical to the 'soul', let alone God). The creative interplay of active matter, imagination, and desire is noted in Pietro Pomponazzi's *On Incantation* (2010). He writes of human composers and singers as *magi* – assisting in the emergence of living simulacra and directing them to material elements and people around them:

> They have in their power the spirit and the blood. In such a situation the imagined and desired thing can be produced in real form by the forces of imagination and desire. The operations of animate things, whether arising from nutritive or sensitive forces, are immediately achieved by the spirit and the blood as if by those forces' own instruments: for the spirit and the blood by their nature are moved by and obey imagination and desire.
>
> (Pomponazzi 2010: 48–49)

Pompanazzi's words resonate with Deleuze and Guattari's materialist philosophy: desire and imagination are real material forces with real material effects. They find echoes in Haraway's comments on the strange complexity of the materiality of nature (1999), which is never adequately framed or tamed by rational and scientific efforts. Haraway reminds us that however committed to materialism we wish to remain, the disciplines of chemistry or biology or physics are 'descriptions' of the world and not 'nature itself'. There is something unutterably more fecund and complex than our models:

> The actors are not all 'us' If the world exists for us as 'nature'; this designates a kind of relationship, an achievement among many actors, not all of them human, not all of them organic, not all of them technological. In its scientific embodiments as well as in other forms, nature is made, but not entirely by humans; it is a co-construction among humans and non-humans. This is a very different vision from the postmodernist observation that all the world is denatured and reproduced in images or replicated in copies.
>
> (Haraway 1992: 297)

Scene zero ...

Operating at the extremes of *strano*, Anna Renzi was not herself. And neither was her audience. Nor need we be. It is not only that *other voices* speak through the body, but also that *'the'* body is a figment just as much as 'the' self or soul, call it what you will. 'The' body is merely the zone most obvious to the dominant senses. We lack senses for the matrices of electromagnetic and chemical and gaseous and even geological conditions required for a single second of any 'human life'. In the overwhelming majority of instances, the eye scans the surfaces, determines a boundary, and says – 'That'll do'. A cognitive penchant for object permanence does the rest. From this, the myths of 'being' and 'human' emerge as if they were truths.

The *Incogniti* were, without doubt, questioning what it was to be human, experimenting with different versions of the body, different versions of the soul: "[They] offered a workplace where they invented 'new thinking', perhaps less as Christians than as free-thinkers, open to new connections and methods" (Chayes 2017: 86). The humorous, seemingly mad, and paradoxical nature of their thoughts and sonic works "do not prevent us from perceiving that this Incognito proposes another way of conceiving the human soul, in debt to Jewish sources [and] the Incogniti elaborated their idea of the soul further into a Kabbalistic mold" (Chayes 2017).

Their theories would confound any surrealist. There was much playful and functional madness, anarchy in the service of emancipation. A key practice was a practice they called "*trasmutatione*, but is known in Judaism as *midrash*. In Talmudic and Kabbalistic writings, this is what scribes do when they 'search out' the meaning of unvocalized Hebrew words and propose variant 'readings' that the absence of vowels easily permits" (Chayes 2017: 85). One *Incognito*, Alessandro Gatti, published a 'scientific' proof that the world is not a sphere, but a cube, using the linguistic juggling techniques of Kabbalism in a long discourse on the ontology of ... cheese.

Renzi's name and reputation give the appearance of one of the first opera stars, but it seems more appropriate to think of her emblem as the infinite Nothingness between the stars. Exploring this operative (operatic) Nothingness has been the aimless aim of our research with an ornamentation methodology (Belgrano and Price 2019– ongoing), allowing a new approach to the emergence of unconscious forces, with no presuppositions about the location or ownership of the forces: they may come from 'inside' the body (memory, desire, etc.), or from the materials the body is working with (air, vibration, a drum beat, a gesture). Creations appear from the interference patterns of two notes – and very possibly many more things of which the participants are unaware. The forces that arrive at some kind of form may come from the vibrant materiality of the world, from the body, from God, or/and Grace, or/and ancestral spirits, demons, from nowhere, from Nothing. The process is epistemologically agnostic. It does not require anyone to believe anything.

> Even the smallest bits of matter are an enormous multitude. Each 'individual' is made up of all possible histories of virtual intra-actions with all Others. Indeterminacy is an un/doing of identity that unsettles the very foundations of non/being [...] Ontological indeterminacy, a radical openness, an infinity of possibilities, is at the core of mattering. How strange that indeterminacy, in its infinite openness, is the condition for the possibility of all structures in their dynamically reconfiguring in/stabilities. Matter in its iterative materialization is a dynamic play of in/determinacy. Matter is never a settled matter. It is always already radically open.
>
> (Barad 2012: 15–16)

Will matter remain open to the Nothingness of spirits and gods such as were vocalized by Renzi? The monstrousness of spirit remains a problem and not just a theoretical one – it is not limited to policing the ontological borders of old or new materialism. The singing voice emerges into the human world, onto the humanistic stage. Then there is the wilderness from which voice comes, and to which it returns. Some things are hardly things at all, very close to Nothing, perhaps even emanations of Nothing. These non-human forces associate themselves with voice and especially with 'pure voice'. This seems to have little to do with terminology. In describing some key features of 'pure voice', we might prefer the short-form term *spirit*: or we might prefer to describe it in terms of energetics and say it is the radical indeterminacy of highly volatilized matter in far-from-equilibrium conditions. Either way, it puts its hooks deep into our systems and gives us the shivers. Some we welcome as aesthetic delights, others are less welcome, depending upon the inter- and intra-actions of the singer, audience, cultural contexts, and the wider environment. Spirit may be Nothing, but Nothing is operative.

Notes

1 http://blog.adw.org/2017/06/sacred-music-history-complex-might-think (accessed 2025-07-15).

2 Josceylin Godwin, https://hermetic.com/godwin/opera-and-the-amorous-initiation (accessed 2025-07-15).

References

Allison, Christine. 2017. The Yazidis. *Oxford Research Encyclopedia of Religion.* Oxford University Press. ISBN: 9780199340378.

Artusi, Giovani. 1600. *Overo Delle Imperfettioni Della Moderna Musica*, The Bibliothèque nationale de France.

Barad, Karen. 2003. Posthumanist Performativity: Toward an Understanding of How Matter Comes to Matter. *Signs: Journal of Women in Culture and Society*. 28 (3): 801–831.

Barad, Karen. 2007. *Meeting the Universe Halfway*, Duke University Press. ISBN: 978-0-8223-3917-5.

Barad, Karen. 2010. Quantum Entanglements and Hauntological Relations. *Derrida Today*. 3: 240–268. https://philpapers.org/rec/BARQEA

Barad, Karen. 2012. *What Is the Measure of Nothingness? Infinity, Virtuality, Justice.* In: 100 Notes, 100 Thoughts: Documenta Series 099. DOCUMENTA (13). Kassel: Hatje Cantz.

Bataille, George. 1990. *Theory of Religion*, Zone Books, M.I.T. ISBN: 10-9780942299090.

Belgrano, Elisabeth L. 2011. *"Lasciatemi morire" o faro "La Finta Pazza": Embodying Vocal Nothingness on Stage in 17th Century Italian and French Operatic Laments and Mad Scenes, Doctoral Dissertation.* Gothenburg: ArtMonitor.

Belgrano, Elisabeth. L. 2016. Vocalizing Nothingness: (Re) Configuring Vocality Inside the Spacetime of Ottavia. *Journal of Interdisciplinary Voice Studies*. 1 (2): 183–195. http://www.intellectbooks.co.uk/journals/view-Journal,id=248/

Berman, M. 1990. *Coming to Our Senses*, Harper Collins. ISBN: 10-004440719X.

Bethell, Richard. 2021. *The Historic Record of Vocal Sound 1690-1829*, NEMA Newsletter.

Burroughs, William Seaward. 2014. *Nova Express* (restored version), Penguin Classics. ISBN: 10-0141396067.

Calcagno, Mauro. 2003. Signifying Nothing: On the Aesthetics of Pure Voice in Early Venetian Opera. *The Journal of Musicology*. 20 (4): 461–497.

Cama, Nicole. 2014. *Defining the Strano: Madness in Renaissance Italy*. https://nicolecama.com.au/project/madness-italy/

Chayes, Evelien. 2017. Crossing Cultures in the Venetian Ghetto. Leone Modena, the Accademia degli Incogniti and Imprese Literature. *Bollettino di italianistica*. 2017–2: 62–88.

Coole, Diana and Samantha Frost. (2010). New Materialisms. ISBN electronic: 978-0-8223-9299-6.

Curtis, Ian and Joy Division. 1979. *Transmission*, Factory Records.

Cusick, Suzanne G. 1993. Gendering Modern Music: Thoughts on the Monteverdi-Artusi Controversy. *Journal of the American Musicological Society*. 46 (1): 1–25. https://doi.org/10.2307/831804

Dolphijn, Rick and Iris Van der Tuin. 2012. *New Materialism: Interviews and Cartographies*, Open Humanities Press. ISBN: 9781607852810.

Ficino, Marsillio. 1489. *De Vita Libri Tres*.

Ficino, Marsillio. 1474/2022. *On the Christian Religion*. Translated by Dan Attrell, Brett Bartlet and David Porreca. University of Toronto Press.

Fisher-Lichte, Erika. 2008. *The Transformative Power of Performance: A New Aesthetics*, Routledge Press.

Ginzberg, Carlo. 1992. *The Cheese and the Worms: The Cosmos of a Sixteenth-Century Miller*, Baltimore: Johns Hopkins University Press.

Ginzberg, Carlo. 2004. *Ecstasies. Deciphering the Witches Sabbath*, University of Chicago Press.

Godwin, Joscelyn. 1979. Layers of Meaning in "The Magic Flute". *The Musical Quarterly*. 65 (4): 471–492. http://www.jstor.org/stable/741568

Godwin, Joscelyn. 1990. *Opera and the Amorous Initiation*, Hermetic Library. https://hermetic.com/godwin/opera-and-the-amorous-initiation

Haraway, Donna. 1992. Promises of Monsters: A Regenerative Politics for Inappropriate/d Others. In: Grossberg, L., Nelson, C., & Treichler, P. (Eds.). *Cultural Studies* (1st ed.). Routledge. https://doi.org/10.4324/9780203699140

Hinton, Peta. 2013. The Quantum Dance and the World's 'Extraordinary Liveliness': Refiguring Corporeal Ethics in Karen Barad's Agential Realism. *Somatechnics*. 3 (1): 169–189.

Laasonen Belgrano, Elisabeth. 2020. Mapping the Burden of Vocality: French 17th Century Vocal Lamentations, Japanese Meditation and Somatic Intra-Action. In: Kapadocha, C. (Ed.). *Somatic Voices in Performance Research*. Routledge Voice Studies: Research Monographs. Introducing chapter: https://www.youtube.com/watch?v=fjUtJ0_xgKs

Laasonen Belgrano, Elisabeth. and Mark Price. 2019–. *No Self Can Tell*, ongoing research project: https://www.researchcatalogue.net/view/980511/988893

Land, Nick. 2011. *Fanged Noumena: Collected Writings 1987-2007*, Urbanomic/Sequence Press. ISBN: 10-095530878X.

Lingis, A. 2000. In: Banham, G. & Blake, C. (Eds.). *Evil Spirits: Demonic Nihilism and the Fate of Modernity*, Manchester University Press. ISBN: 10-071905642X.

Manzini, Luigi *Il Niente*, Venice, 1634. In: Ossola, Carlo (Ed.). *Le Antiche Memorie del Nulla*, Rome: Edizioni di Storia e Letteratura.

Marlan, Stanton. 2008. *The Black Sun: The Alchemy and Art of Darkness*, Texas A & M University Press.

Marsden, Jill. 2020. Poetic Connections: Sympathy and Community in Whitman's "Song of Myself". *Mosaic*. 53 (3).

Miato, Monica. 1998. *L'Accademia degli Incogniti di Giovan Francesco Loredan, Venezia (1630- 1661)*. Firenze: Olschki.

Mozart, Wolfgang Amadeus, Johann Baptist Henneberg, Benedikt Schack, Franz Xaver Gerl and Emanuel Schikaneder. 1790. *The Philosopher's Stone, or the Enchanted Isle*. Singspiel in two acts. Vienna: Libretto.

Nietzsche, Friedrich. 1889/1997. *Twilight of the Idols*, Indianapolis: Hackett Publishing Company, Inc.

Pomponazzi, Pietro. 2010. *On Incantations*. Translated by Lynn Thorndyke. Kessinger Publishing, LLC.

Pope, Charles. 2017. *What is Sacred Music? The History Is More Complex Than You Might Think*. http://blog.adw.org/2017/06/sacred-music-history-complex-might-think

Rig Veda, Book VII, Chapter 104, Verses 1, 10, 11.

Ray, Meredith. K. and K. J. P. Lowe 2015. *Daughters of Alchemy*, Harvard University Press. http://www.jstor.org/stable/j.ctt1c84d90

Roob, Alexander. 2014. *Alchemy & Mysticism*, Taschen. ISBN: 978-3836549363.

Rosand, Ellen. 1991. *Opera in Seventeenth-Century Venice: The Creation of a Genre*, University of California Press.

Ruggiero, Guido. 1978. Law and Punishment in Early Renaissance Venice. *Journal of Criminal Law and Criminology*. 69: 243. https://scholarlycommons.law.northwestern.edu/jclc/vol69/iss2/8

Strozzi, Giulio. 1644. *Le Glorie Della Signora Anna Renzi Romana*, Venice: Surian.

Van der Berk, Tjeu. 2004. *Die Zauberflöte. An Alchemical Allegory*, Brill.

2 The onto-ethico-aesthetic force of musicality

Jannie Pranger

> My performances ... conveyed the impression that a work was created at that moment .... Though they were performances of pieces I knew well, the template had been forgotten .... [T]hey were distinctly, exaggeratedly material, directing attention to the physical reality of the musicians, the sounds ... and one's place as a listener or performer within that sound.
>
> (Carolyn Abbate, *In Search of Opera*, 2001: xv–xvi)

In this chapter, I engage with singing as a practice via which to approach a problem facing the musical event, namely, its lively musico-material fleshy responsiveness—that of touching, and of being touched.[1] Singing demands one to actively feel and respond to the historicity of a marked vocal body; a lively body of heterogeneous, interdependent musical forms and materiality, trained to participate in and anticipate the production of tone, pitch, loudness configurations, and verbal articulations (Tiainen 2007).[2] Furthermore, singing is a demonstration of responsiveness to the social forces at play, requiring responsiveness to the working of an acoustic surrounding and (when used) of electronic equipment (e.g., a microphone, sound reinforcement, or recording system). In short, the activity of singing opens itself to multiple musico-material encounters that produce a relational dynamic through and within which a musical event and a vocal body become temporally and locally (re)configured.

When analyzing vocal practices, we commonly see ourselves confronted with a musico-material and discursive constitution—a bounded practice with defined musical values and defined material properties—enabling judgments to be made both about doing research and, as a vocalist, about making music and about how to sing. Evidently, a vocalist's participation within a rock, jazz, folk, hip-hop, or classical music scene has consequences, not only for wider musico-material boundaries and properties, but also for the vocalist's own skills and other performance competences.[3] My analysis, however, starts neither from a given musico-material vocal practice, nor does it address the plenitude of possible musical genres in relation to the abundant capacities of a vocal body (which again would require starting from defined

DOI: 10.4324/9781003390992-4

musico-material categories, the underlying criteria of which enable differentiated musical genres, possibilities, capacities, and vocal bodies). Rather, my aim is to explore a felt complexity, "always already embraced," that traverses the regulative norms of constituted musical practices (Vicki Kirby paraphrased in Tuin van der 2013: 7). Although the boundary working of, for example, a classical musico-material practice may appear to remain intact, while singing, the trained vocal body becomes one with dynamic musical and material forces, open to processes of invention and creation. Here, we are faced with the complexity of touching and testing through and within which a vocal body and the sonorous gesture become made and re-made, suggesting that a vocal body and the sonority produced are engaged in a confusing play of connecting and disconnecting, carried away from the musico-material norms of a specific practice that are necessarily being adopted.[4]

The quote of musicologist Carolyn Abbate, with which I open above, addresses this confusion. Abbate (2001) troubles the representationalism at work in the assumed correspondence between a template—a compositional pattern or underlying structure—and a vocal gesture.[5] As a trained listener, Abbate witnesses how a vocalist is not engaged in following a template, singing the "same" piece, the "same" note, or the "same" word. Rather, elaborating on Abbate's problem, a vocalist is taken up in a process of ongoing differentiation through and within which template, musical gesture, and a vocal body—and their relationality—materialize. As such, Abbate's statement might be widened to the difficulty of a vocalist who is simultaneously dislocated by being caught up in material forces *and* located within the boundaries and properties of a given musico-material practice. In this broader sense, the philosopher and feminist theoretician Rosi Braidotti (2012) supports Abbate's experiences when arguing how "the [Deleuzian] notion of the univocity of Being or single matter positions difference as a verb or process of becoming at the heart of the matter" (28). By suggesting that difference in terms of a difference between entities is traversed by generative processes of *differing*, this calls for thinking musical becoming from the act. This necessarily entails not starting from the defined, individual element (in which case material or cultural regulative forces would act as both the external ends and the means for the production of difference). This leaves us with the intriguing puzzle of how musical and material normativity, when viewed as a difference that makes a difference, emerges within and through univocal Being.

Singing, as I hope to demonstrate, harbors this musical and ontological puzzle, redirecting our engagement with musical aesthetics and ethics as a touching and testing at the heart of vocal becoming. In my exploration, I engage with this musico-ontological problem while considering its musicological relevance. This guides my inquiry into how singing's lively responsiveness might trouble the common concept of musicality. Musicality (and with it, unmusicality) is a powerful concept linked to what is mostly understood as cultural operations that connect musical aesthetics to bodily and material activity. Troubling musicality means then confronting its notorious dualist schema of pre-definition—based on a pre-defined relationality between a pre-defined concept of music and the musico-material act—with processes of differing that affirm the inseparability of the musical and the material.

Briefly turning to the common notion of musicality, we find a concept of difference as an alterity between musical value and human properties. In these terms, musicality is predicated upon prior judgments, enabling one to decide whether human properties do or do not have the quality of a particular music value. Or, in more familiar terms, musicality is considered a particular human skill and talent related to a recognized concept of music. Also, academia has studied the concept of musicality in terms of a pre-defined relationship between music and musical practice. Although musicality has been studied from many different angles—the music-sociological, cognitive, and biological—the emphasis and starting point of such research is based on an understanding of the difference in which one of the terms—musical value or material doing—regulates the other (as detailed below). For now, what might be asserted is that common understandings, both vernacular and academic, build upon pre-defined concepts of music, musicality, and their relationship, in which musicality is seen as the universal (human) capacity to respond to recognized music.

Yet, when taking the vocalist as our guide, we cannot help but encounter the inherent "liveliness" of entangled musico-material responsiveness, which complicates musicality as a pre-defined quality involving an exclusively human and innate capacity of musical intention. My exploration thus concerns what musicality can be when conceived as a responsive "doing" generated by a differential process of musico-material creation—one that suggests that the "musical" and the "material" are produced within and through each other. As such, my exploration of musicality subscribes to a new materialist ethos and its concern to *re*-think and *re*-search the dynamic and generative relationship of form and matter (Dolphijn and van der Tuin 2012: 91). I especially build on feminist research and theorizing for which the materiality of bodies, as well as their "material-semiotic" (Haraway 1988: 595) and "material-discursive" (Barad 2003: 810) nature, has long been a productive source (Alaimo and Hekman 2008: 1). In this manner, I affirm how, in feminist research and theorizing, new materialism, as a general term, indicates a renewed attention to the *mutual constitutive* relationship between nature and culture. This feminist materialism challenges the anthropocentric practice of humanism, which suggests that knowledge, being, and value are predicated upon the "human [observer] in its Renaissance iteration"—that is, taking the human as the measure of knowing, being, and valuing to the exclusion of all Others (Panurgiá 2018: 44).

Relevant to my explorations are the insights of the physicist and philosopher Karen Barad. Her notion of intra-action, understood as entangled encounters, supports my reworking of musicality, bringing ethics and aesthetics into play as the non-transcendent testing and touching of emergent (musico-material) Others. This should immediately make clear that this chapter does not aim to establish a new theory of musicality. Rather, its aim is to open up the concept of musicality, considering it to be a musician's and musicologist's problem of entangled, musico-material responsiveness and a creative politics of musico-material constitution. My hope is that, instead of the cultural, the social, or the material acting as a punctum from which to start, the reworked concept of musicality will enable a music/ological engagement—of knowing, doing, and valuing in being—with the entangled and performative nature of musico-material practices. My guide will be the vocalist's practice without necessarily defining the musico-material boundaries and properties of that practice. In my reworking, musicality as a responsive touching and testing, always already embraced and felt, alludes to a more intimate relationship with the musico-material forces at stake. This musicality contains the puzzle of musico-materialization in which the constitutive and the constituted cannot be segregated and do not pre-exist as such their relationality.

I will start with an examination of commonly held beliefs regarding musicality and how those beliefs work through in research and theorizing. An important part of reworking the concept of musicality will involve an engagement with Barad's understanding of relational ontology. Therefore, I follow with a more thorough discussion of Barad's conceptualization of difference as an effect that is generated through and within intra-activity. Finally, I will suggest possibilities that arise if musicality and unmusicality are considered musico-material relational "doings" and what such a move might offer to the musicological study of musico-material creative invention.

Toward a generated and generative musicality

In a contemporary context, the concept of music and musicality forms a dualism, the properties and boundaries of which are considered to pre-exist their relationship. Musicality implies a particular state of being, referring to a particular set of human talents, skills, and behaviors—for example, the musicality of the performer and the musicality of speech. Notwithstanding the inherent ambiguity of the noun "musicality" and the adjective "musical," and offering a degree of orientation on a notion of what music is, I would nevertheless suggest that the concept of musicality is known as referring to inherent properties (e.g., the gifted musician) or learnt properties added to humans and things, which are related to, and pre-exist, a general concept of music as a social and aesthetic "given." This suggests a relationship between musicality and music that implies at least a time order in which music and musicality seem to compete in pre-existing each other. Is musicality, as an added value, the echo of music that is yet to come, or is perhaps the converse true? Although a time order—from past to future, or the converse—may trouble time's parametric steadiness, it fails to account for time as an operator, which would affirm a musico-material historicity. In other words, contemporary notions of the relationship between music and musicality may stir up the ordering of past and future events, without challenging the material, the social, and the aesthetic as given points on a timeline, and expressed as inherent and added human musical properties.

Yet, and without critiquing the scholars and scientists mentioned below, this dualist concept of musicality (and music) pervades arguments on the subject of musicality (albeit often without being noticed) by most commonly referring to the specific qualities that are "owned" by humans, animals, or things, in relation to performing, dancing, and listening with an overarching general notion of pre-existent music (Théberge 1997: 172; de Mink 2013). Others maintain a dualist and human-centered notion of music and musicality in terms of the human capacity to be musical (Martinez and Villanueva 2018); of human sense-making in a defined musical environment (Ojala 2014); of music education (Knieter 1967; Toporova et al. 2016); or of studying musicality as a concept that enables us to compare human behavior across populations (O'Flynn 2005; Trehub et al. 2015). The following two studies address the concept of musicality itself—both its performative qualities and as an innate capacity to connect to a musical environment.

The musicologist Philip Brett (2006), when accounting for its social complexity, considers musicality to be an ethical and aesthetic problem of inclusion and exclusion. In Brett's argument, musicality becomes a regulatory force based on an ideal concept of music—one instrumental in a mechanism of social power. Brett discusses musicality as a label and a "tool for social control dressed up ... as talent," which, in relation to its being an ideology, supports "a concept of 'art' as an elite and segregated human activity" (22). Brett's main concern points to feminist philosopher and gender studies scholar Judith Butler's notion of power as a performative structure in a social domain—how a concept of art, distributed as a power structure in a musical world, "[produces] the subjects [it] subsequently [comes] to represent" (Butler cited in Barad 2003: 804).

We find musicality as an object of research in itself in the work of cognitive musicologist Henkjan Honing et al. (2015). They define musicality and its relationship to music as a "natural, spontaneously developing trait based on and constrained by biology and cognition" (1). Music, according to them, "can be defined as a social and cultural construct based on that very musicality" (1). Their argument considers musicality to be a specific "innate" aptitude of humans and animals, enabling a particular orientation to an environment—the seat from which music as a social and cultural event has evolved (3). These musicologists, all with a general interest in the evolution of music, seek the biological and cognitive components of which musicality is assumed to consist (3). This approach guides their intellectual interventions, privileging questions on the biological and cognitive constitution of musicality. Musicality here tends to be considered an *a priori* biological capacity to "act musical" in terms of cognition—that is, knowing, perceiving, and appreciating, with the implication of recognizing that which is musical (Honing et al. 2015).

The studies of Brett (2006) and Honing et al. (2015) posit musicality as a property (respectively, added and innate) of individuals brought into play through the working of social power structures or human (and animal) perception. Both studies affirm the constituent aspects of musicality in tandem with regulative principles. Yet, for them, musicality is a known constant that may vary according to social structures (Brett 2006) or structures of biology and human (and animal) cognition as marker of consciousness (Honing et al. 2015). In my reading, acknowledging how my intervention privileges the problem of an entangled musico-material dynamic, Brett (2006) and Honing et al. (2015) overlook musicality as an emergent responsiveness *becoming normative* from within, both generated by and generative of encounters in the flesh. In other words, to open up the musico-material performative and generative dynamics of musicality to thinking necessarily implies a re-questioning of the boundaries of the human, the social, and material life, and a re-questioning of the values that constitute them. To emphasize the fundamental shift in thinking to which these questions point, I will use a limited palette consisting of a vocalist in the midst of dynamic musico-material being, the insights of Barad, and my own experiences as a vocalist.

Considering the research mentioned above, musicality is embedded in a "matter-of-factness" and carries with it real value in a real world of human (and animal) bodies—and in their real social and material environments. However, questions arise when taking account of the vocalist in the midst of singing. What if this matter-of-factness is riddled with ambiguity, where "[o]ntological indeterminacy, a radical openness, an infinity of possibilities, is at the core of [musico-] mattering" (Barad 2012b: 214)? Here my limited palette is of help, imagining and bringing to the fore the inherent ambiguity of other questions prompted by the musico material dynamics of singing: What if a vocal body and an acoustic environment both feel and think? What if musico-material *intra*-activity, as detailed below, is a general occurrence that simultaneously produces a vocal body and a sonorous gesture with utmost precision, undermining human intentionality? What if unmusicality is not a label for denigration but means facing the "indeterminate non/being non/becoming of mattering and not mattering," enabling ethics and aesthetics of musicality to arise (Barad 2012b: 216, emphasis removed)? What if there is no "before" and no "original" foundation of musicality—that musicality is *of* the sonorous world?

Such questions suggest a musicality open to experimentation and the unforeseen, grappling with modes of materialized responsivity such as touch (e.g., a vocalist's breathing), care (e.g., the durational effort of staying with the tone), and supported by a desire for the sonorous. In short, what if we go along with the vocalist's invention in the midst of singing, troubling a musicality that judges the musical act against a pre-defined concept of music? I engage with these questions using the insights provided by Barad, specifically Barad's (2007) concept of "intra-action," which I read as the crystallization of the elaboration of a relational ontology.

Relational ontology: A metaphysics from within

Commencing her inquiry from an entangled onto-epistemology, Barad (2007) nullifies any thinking that starts from the premise of isolating being and knowing, thus challenging existing research and theorizing that start from fixed object and subject positions with an implicated hierarchy in which a subject is assumed to take an active role against an object. In the musical field, we encounter a similar dynamic when a vocalist assumes herself to be the source of musical intentionality set against a vocal body, here seen as an obstacle that needs coordination and must be animated by breathing musical life into it. From an academic perspective, Barad's non-dualist approach challenges precisely this assumed gap—as elaborated by Carolyn Abbate (2004)—between the musical event as an object of study and the knowledge-seeking musicologist. Both examples—the human as a source of isolated intentionality and as an isolated knowing subject—allude to the age-old struggle with the problems of access to the (musico-)material world. Supported by the thesis that "we are *of* the world"—that we attempt to understand and help to create—and via her onto-epistemological approach, Barad overcomes this issue of access, suggesting that we, as human mind-bodies, are always already part of the world's ongoing (re)configurations (Barad 2007: 185, emphasis in original).

The question of Barad's (2007) onto-epistemology and her notion of immanence suggest that musico-material-discursive engagements are inseparable from the bigger picture of being part of the world's ongoing transformations. The perplexing aspect, however, is not so much that immanence may announce the redundancy of access issues. Rather, what we need to consider are the implied methodological questions, such as how knowledge is possible when one is an "inseparable part of the world," and subject and object positions are not pre-defined. Here, I think of the ethico-aesthetic questions that arise when we face up to the real consequences of music that is not a context in which we live, play, sing, and study, but is an inseparable part of a sonic universe from which we, our thinking, and doing come into being. To unpack this further, I turn to those insights of Barad that relate to my exploration of musicality and unmusicality as emergent calls from a musico-material practice of singing.

Barad's (2007) research emphasizes the performative nature of scientific practices, pointing beyond the typical scientific objectives used to demonstrate her ideas and, as such, positively feeding inquiries into the dynamic and performative nature of musico-material practices.[6] However, caution is in order. Barad's work should not be read as a handbook or framework of general principles concerning the world's ongoing creativity. Rather, recalling its methodological implications, one should consider how her work does not set feminist and physics practices against each other but rather is generated from their relationality. Nevertheless, the lure of simply translating Barad's insights into musico-material matters is made palpable in the manner in which Barad (to my knowledge, only once) addresses the arts in *What Is the Measure of Nothingness? Infinity, Virtuality, Justice* (Barad 2012c) written on the occasion of the dOCUMENTA 13 exhibition in Kassel (Germany). Barad explores the world's ongoing creativity as demonstrated by a quantum field theoretical perspective on the void's material liveliness—a grand and inspiring gesture for artists and art lovers alike that asks how a quantum world perpetually surprises and renews itself. However, for this inspiring gesture of ongoing creativity to thrive materially and meaningfully in studios, in rehearsal rooms, on stages, and in musicological quarters, an encounter as a critical practice of *engagement* is needed. Here again, the methodological question arises. Yet, starting from within vocal practices, it is helpful to keep in mind how we are already familiar with such a critical practice of engagement: I simply refer to *rehearsing* in all its musico-material complexity. Via a process of testing, touching, doing, and re-doing, a sonorous gesture and vocal body are mutually constituted into song.

In terms of knowledge practices, Barad (2003, 2007) elaborates upon such engagements as methods of "diffraction," referring to the physical phenomenon and to the experimental set-up for research into the nature of light.[7] In short, a diffractive approach starts by reading insights *through* one another (the experimental set-up). From this activity agencies of research (i.e., both researcher and experimental set-ups, as well as an object of research) are generated—the diffracted effect of difference.[8] Here, it is crucial that such a diffractive approach traverses the dialectic of reading pre-defined things or thoughts as posited against one another, but rather is concerned with the generative effect of difference when insights are diffracted—just as a rehearsal thrives, not on the difference between the vocal body and sonority, but on diffraction as a generative relationality and on the song as an effect of that relationality. In this sense, reworking the concept of musicality is an effect of "reading/thinking/listening" together with Barad's insights, musicological studies, and my intuited (materialized) knowledge of singing. Diffraction (or rehearsal) is thus a method of research that includes the researcher—one that cuts through boundaries and thrives on curiosity and alertness to the generative effects of the encounter.

In the context of diffraction, Barad affirms a relational ontology and performative account of practices in which materiality is not a fixed substance in nature. Rather, it is a "doing" that concerns nature's "ongoing historicity" (Barad 2003: 820–821). This means that coming to matter does not require the mark of an external force. Barad (2007) argues that meaning and intelligibility are not "mind-based capacities, [but concern] marked historialities ingrained in [musico-material] becoming" (393).[9]

Digital music studies scholar Aden Evens' (2005) notion of "suitability" precisely grasps Barad's notion of material coming to matter as matter-meaning formation, not involving human cognizance.[10] For example, the intra-active encounter between sound waves produces a particular responsivity that matters materially, producing a qualitative change. Whether colliding waves cancel each other out when out of phase or reinforce each other when in phase, the effect, nonetheless, is a new configuration with real effects. We might say that through and within the coming together of waves, a felt mutual suitability matters to the timbre and loudness of sounds as the temporal effects of an included marked historicity of the concert hall acoustics and the vibrations issued from an included marked historicity of a vocal body. Indeed, practices are entangled musico-material doings, constitutive of matter-form relations that are profoundly undefinable in human terms as to where materiality ends and music or meaning begins. Think of how such musico-material formations work through in musico-material practices. For example, the relationality of a microphone and a vocal body implicates a notion of suitability and iterative dynamic of new matter-meaning configurations. The specific microphone technology orients a vocal body's thinking and actions, which orients further exploration of microphone technologies, which orients a vocal body's thinking and actions, and so forth (Lockheart 2003; Eidsheim 2015). Also, a body's materiality demands the mutual suitability of a vocalist's breathing and its biological processes—if not, the felt effect could simply be dizziness. In short, matter-matter encounters suggest a material liveliness that includes meaning formation, producing a relational dynamic that complicates the stability of musico-material practices as a *relative* stability, as well as opening up their regulative musico-material norms to change.

Furthermore, music as discursivity does not involve linguistic (and symbolic) representations that would function as a stand-in for a musico-material world. Instead, representations are made and involve specific musico-material-discursive practices of (re)configuration, producing "local determinations of boundaries, properties, and meanings," and hence, the boundaries and properties of a *specific* musico-material or musicological practice (Barad 2003: 820–821). In Baradian terms, concepts and other symbolic means are not general abstractions that capture a material world, nor are they written into a material world, which would suggest that the world is produced by human concepts.[11] Concepts and other symbolic means do not stand on their own but rather are "material enactments that *contribute to*, and are *part of*, the phenomena we describe" and that result in, not only the notion of musico-materiality (Barad 2007: 32, emphasis added), but also the necessary reworking of music, musicality, and their relationship in the light of a vocalist caught up in dynamic musico-material doings, suggesting a more intimate relationality between doing, valuing, and event.

The remaining concept relevant for my exploration is Barad's (2007) understanding of intra-action. Intra-action crystallizes previous notions of diffraction and the double movement of musico-material encounters, involving the inseparability of musical and material expressive forces. Intra-action concerns the "*mutual constitution of entangled agencies*" (33, emphasis in original)—in the context of our investigation, agency, as the becoming precise of musico-material actions, is an emergent effect of musico-material encounters. Here, precision entails how, through and within intra-activity, agency, cause-and-effect relation, and subject and object distinction emerge, effecting a particular change that opens to further intra-activity. Intra-action thus underpins Barad's relational ontology; the world does not come into existence via discrete objects that interact, but subjects and objects (their agencies, properties, and boundaries) emerge *because* of their intra-action. In other words, instead of inter-action—a relation between determinate, individual entities, and localities assumed to pre-exist the encounter, maintaining a degree of independency—intra-action involves the coming to matter, musically and meaningfully, through and from within the meeting-action. In this way, intra-action reveals how each encounter results in specific materializations, displaying how differences emerge, also suggesting a notion of how the world is caught up in a generative relationality—a "working through" in every fiber of its being (Barad 2007).

Taking up the question of musicality, these volatile yet precise intra-active encounters bring into focus how musico-material practices that constitute the relative stability of the "this and that" of "our" music and of "singing on the beat" are, in Barad's (2015) terms, "threaded through with an infinite alterity" (401). This suggests that intra-actions introduce, through and from within musico-materializations, the unforeseeable from which invention comes into being, contesting any limit of musicality and potentially destabilizing the musical and material conventions that distinguish a particular musical practice. That is, considering relative stability as the grounding and ungrounding dynamics of a relational ontology, a politics of testing and touching is introduced when things are not pre-defined and bound to so-called "natural" processes of unfolding. For example, when considering music as the stable background against which we sing and study, we might think of how a song unfolds—is composed, learned, memorized, performed, recorded, distributed—as a succession of musical events. However, by shifting such a dualist frame of thought and leaping into the sonic universe, we are caught up by an ongoing dynamic of material and musical expressiveness that calls for touching and testing, whether we sing, research, or theorize. In other words, rather than referring to transcendental truths, aesthetics *and* ethics then become an inseparable part of mattering. Thus, if ethics is not a matter of how we ought to behave as against the Other and aesthetics is not a matter of how a musician ought to perform as against a concept of (musical) beauty, both defying the generalization of ethical and aesthetical concerns, then the notions of ethics and aesthetics might converge in a concept of musicality that opens to questions of *how* musico-matter comes to *matter*—for whom and for what. Musicality thus becomes a problem of experimentation and musico-material responsiveness, demanding alertness and sensitivity to an ethico-aesthetic call from within musico-material "doings."

Musicality and intra-activity

Thinking together Barad's notion of intra-activity and an intuited (musico-materialized) knowledge of the ontological complexity embraced by the vocalist paves the way for a concept of musicality that is more necessary, intimate, and profound—one neither instigated by an ideal aesthetics, *the* social, or *the* material, nor by an exclusive human property to act appropriately. Put simply, while singing, a vocalist *must* stay with a tone, a necessary effort that accompanies the leap into musico-material dynamics, articulating a desire for the sonorous. Searching music on the deepest level of musico-material relations, a vocalist becomes engaged in an intimate relationality in which she emerges as a vocal body together with the musico-material object encountered. While singing, a pitch crystallizes as *that* F, or a vowel crystallizes as *that* sonority. In both cases, singing becomes the ontology of *that* musico-material object and inseparably related to *that* vocal body. In other words, the vocalist's "doings" elicit a deeper notion of musico-materialization as a relational ontology, beyond music that is embodied or incarnate or, conversely, beyond a musically embodied sound. The mutual intra-active sensitivity of a musico-material trained body and an acoustically eventful environment makes that music and matter lose their character as social, cultural, or material inscribers. Instead, they become inseparable, entangled forces in musico-materializations. In short, demanding a more profound *intra-active* concept of musicality as a generative and relational "doing" from the musico-material event.

However, a difficulty emerges here, namely that such an intra-actively produced musicality, through and within dynamic musico-material processes, can, in itself, not be thought. Henri Bergson ([1889] 1910) elaborated the concept *durée* (duration) to indicate how such forces of dynamic being cannot be taken up by the spatializing and generalizing tendency of the intellect's explanatory power. What *can* be done is to map the constituting effects of musicality that mark the creative processes of a musico-material world. Marks, in this sense, are not the outlines of a processual temporality or, similarly, set points on a fixed map, but rather concern intuited and felt moments with bodily effects that come into being within and through the relational dynamics of musico-materializations. A mapping act, in this sense, touches a real and alive musico-mattering—the felt and the intuited—both materially and meaningfully. Here, I elaborate upon the Baradian concept of intra-action to "refer to a nonconceptual [or ...] intuitive understanding" (Deleuze and Guattari 1994: 40) of musico-material invention and creation *in the event*.[12] In this sense, intra-action involves how qualitative difference emerges through and from within entangled encounters, enabling us to reconceptualize the role musicality plays as a mode of inventive and creative "doing" (re)orienting musico-materializations.

As a personal account (which, I dare say, would be confirmed by other performers), during the musico-material turbulence of a concert, there are moments in which I feel a strange clarity. Certainty or uncertainty about the justness of breath, bodily coordination, the production of a tone, a timing, or of the anticipation of an event that "unmistakably" seems to await in the accompaniment. From a Baradian perspective, such moments of clarity mark a generative relationality, suggesting what Barad (2007) calls a *cut* of intra-active in/separability as an "exteriority within" (135). This means that a difference—between a vocalist and a musico-material object—is constituted without undoing the relationality of the vocalist's and the object's entangled being in musico-materialization. Rather, their *intra*-active in/separability constitutes a resolution through which ontological indeterminacy comes to a defined musico-materialization (Barad 2007: 140). The flash of felt clarity is then an effect of an intra-active move in which a cause-and-effect relation, musicality as precise musico-material doing, and a subject-object distinction come into being, articulating how difference that makes a difference emerges from the act of musico-materialization (Barad 2007: 389–391). What these felt moments of musico-material "truth" demonstrate is how the relational doing of intra-actions, orienting musico-materializations, involves musicality as an ongoing force that potentially undermines the primacy of a vocalist's musical plan for a concert and, in a wider sense, the gravity of musical and material conditions of vocal practices.

The moment of "truth" during a vocalist's affective engagement with un/expected musico-material events also means that he or she witnesses that something comes to musico-materialization. However, as suggested earlier, singing's ethicality and aesthetics do not point to a cognized matter-of-factness, judged against some yardstick enabling a "yes" or a "no." Rather, the resolution of an indeterminate intra-active in/separability brings to the fore a relational precision of *how* something matters or not—for what, for whom, and how it orients future musico-materializations. In other words, guided by a vocalist, we see that the vocalist, instead of judging musical truth on pre-defined conditions, becomes involved in experimentation not only through musico-material touching, but also by being touched, and "being there." The same can be said of ethics; instead of judging pre-defined good and just behavior toward a pre-defined other, it takes the form of musico-material experimentation with real consequences that have an effect on all that is intra-actively called into being. This means that a "truth" involves a felt mark of musico-materialization, constituted through and within intra-actions, not along the lines of judgment and critique, but rather along affirmative lines enacted in processes of musico-material invention and creation.

Musicality enters through and from within intra-active musico-materializations as responsiveness—built on musico-material alertness and a desire for the sonorous, expressed through testing and touching—that grows into precise actions, inducing change as a difference that makes a difference. As such, musicality becomes an onto-ethico-aesthetic force of normative, differential responsiveness entangled with the specific musico-material intra-activity from which it emerges. The intra-active character suggests that normativity does not appeal to particular pre-defined regularities, such as the correct response to a stimulus (for example, correctly singing the notated F, or remembering the musical phrase correctly). Nor does the term "differential" suggest the possibility of responding differently to different musico-material stimuli, as *inter*-action would suggest—for example, recalling the notion of skill and competence, how a singer has learnt to respond differently to different acoustic circumstances. Rather, musicality as a normative responsiveness is an emergent effect of intra-actions in which different musico-material encounters produce different modes of musicality, not only resulting in different musico-materializations, but also having real effects on new intra-actions.

Musicality: An investment in ongoing variation

As a normative, differential responsiveness, musicality affirms the musico-material liveliness that a vocalist in the midst of singing embraces. Every intra-activity generates new musicalities, which cannot be predicted on the basis of the musico-material formation of a particular vocal practice, the structure of a composition, or that which is learnt by heart. In other words, musicality does not involve the capability of fusing together elements—the rhythms, harmonies, pitches, and words—of a song into a melodious whole. Rather, musico-materialization, in terms of the relative stability of a song as a "finite" event, might be considered a condensation of musicalities that orient an unforeseen potential of *marked* musico-materialities toward an unpredictable future. Marked means that a song's elements have a historicity of determined musico-material properties, but being taken up in new intra-active engagements opens to their potential variability. For example, multiple musicalities are engaged in the coming together of a vocal body's actions and the gravitational force of a conductor's arm movements, resulting in the intensity of the *now* of *this* sound. This suggests that a "finite" musical event involves a temporal, local, and dynamic condensation of musicalities that flow from the desire for a better sonority-to-come—that is, from a value-laden touching and testing, oriented toward a future of what music could be. This implies that there are no defined elements, no vocalist—indeed, no ideal music—behind musicality and vice versa. Rather, musicality is the intra-active engagement *from* which a vocalist and musico-material Others are generated as creative participants in musico-materializations, suggesting multiple musico-material centers of condensed musicalities endlessly made and unmade, both locally and temporally.

To sketch the scope and shift in thinking required of such musicality, we might think how intra-active musico-materialization—as resistant to pre-defined reference points or centers—envelops centers such as those presupposed by Copernican revolutions. In their different expressions, such revolutions in thought still start from some central point—be that earth, God, or the human subject—around which they rotate or to which they refer (Deleuze [1994] 2004: 201). In the insights of Barad and Deleuze, we find a thinking-through of the world's liveliness as an ongoing differentiation with temporal stabilizations of multiple centers that are made and unmade in a dynamic field of ongoing variation. Clearly, the scope on which the analyses of Barad and Deleuze operate—involving a relational, ontological dynamism traversing quantum, earthly, and cosmic matters—might seem dizzying. However, guided by a vocalist in the midst of musico-material (re) configurations, we find a musicality that, indeed, intra-actively operates in all directions open to variation.

Here, there are two things that may offer protection from latent dizziness. Following the thesis of immanence—the idea that we are an inseparable part of the sonic world that we attempt to understand and help to create—singing and thinking are generated by, and generative of, musico-material practices and do not transcend the totality of musico-material matters. Furthermore, a dynamic, sonic universe of ongoing variation and musicality that operates in all directions does not suggest a metaphorical fluidity that flows in time and space, where time and space are given factors transcending musico-materializations. Rather, a musicality that intra-actively operates in all directions involves a new way of thinking relationality—with each other, with matter, materials, musico-material practices, and discourses. Every musico-material intra-action is potentially constitutive of a cascade of musicalities, producing change or transformations, for example, in the movement of air and climate of an acoustic space, in the historiality of musico-material practices, in the "doings" of listeners and performers across times and spaces, and related through technologies of recording and distribution. These are not idealized movements but rather concern real musico-material "doings" with real effects, demonstrating the generative role of musicality in musico-material "becoming" (Deleuze and Guattari [1980] 2004) or "[musico]spacetimemattering" (Barad 2007).[13]

Imagine how an audience and a vocalist are brought together in a concert hall at a particular date and time—one that intra-actively relates to their agendas, their locations, and their forms of transport. Yet, also think how this given locale—its time and place (space)—becomes enfolded in musico-materializations, in which emergent modes of musicality determine vocalists, listeners, sounding gestures, and their relationships. In a wider sense, this emergent musicality is implicated in the world's intra-active and sonic dynamism. As such, it may involve an ongoing material touch, causing a disturbance in air, water, or solid matter, taken up in musico-materializations, producing rhythms and melodies through and within intra-active responsive tapping, rubbing, striking, breathing, and so forth. In other words, by thinking of invention and creativity from relationality, intra-actively performing strange leaps, musicality becomes the onto-ethico-aesthetic expressive force of musico-materialization.

The question of unmusicality

In this final section, I explore what it means to enter into an intra-active musico-material relationality, emphasizing in musicality the tendency to orientate itself toward a future as a yearning toward an unheard, imagined sonority. This means an orientation to what comes *from* the future, entailing a testing and touching of potentiality from within the dynamics of heterogeneous musico-materializations. This calls for detailed observation, suggesting a crucial distinction from how singing or listening to a song's common unfolding of events enables a vocalist and listener to project a meter-rhythmic or chord-pitch progression into the future.

Once more, we find the vocalist in the midst of singing, now surfing the crest of musico-material becoming, at the edge of in/determinacy, and before an intra-active resolution becomes defined, resulting in a particular musicality and, with it, a particular musico-materialization. The vocalist is confronted with the perplexing nature of multiple material doings, accumulating im/possible musicalities that inhabit the before of coming into a musico-material intra-active encounter, before a breath enters into musico-material intra-action. In other words, before breathing finds its ground through an intra-actively produced musicality, constitutive of a specific inhalation, which already embraces the imagined—sensing the insensible—tone of the exhalation: "Inhale the vocality-to-come!" said my vocal coach. The vocalist, awake to musico-material doings, involves the "forefeeling," not of a defined sonority, but of an imagined, unheard sonority that accompanies a state of in/determinacy, calling upon the vocalist in the intense moment of the "not yet."

The effort of sensing the unheard, indeterminate, yet yearned-for sonority, widens musicality as the intra-actively produced normative differential responsiveness, to include an onto-ethico-aesthetic force from the future. This suggests that the yearned for, imagined, yet nonetheless insensible sonority has a real effect in the complex moment in which the "how" of musico-materialization begins to take shape. This yearned-for, imagined sonority demands one to be alive to the Baradian slash, troubling the absolute separation between past and future occurrences, creating an entangled temporality and dis/continuous musico-materialization—one captured, neither by the repeatability of thoughts nor by the actions of the past into the future.[14] Think of how a vocalist—also when singing from a score, when trained in a specific musical tradition, or when following a conductor—does not exactly follow a model in which training and musical capabilities would make the vocalist fit for the task. While singing, vocalists do things that they never did before during their training, rehearsals, or concerts, suggesting that the intra-active dynamics of musicality involves responses that are situated, material, simultaneously meaningful *and* unforeseen. Such intra-active generated musicality, being sensitive to an imagined sonority, may evoke the lure of responsivity barely finding words ("… oops! … to late! … yes! …") and reaches out with expectation ("… yes! … there's my timing!"). What these felt and intuitive moments tell us is that thinking with a musicality infused with sensing the unheard enables us to think the intra-active simultaneity of an imagined, insensible sonority and its real intra-active effects within heterogeneous musico-materializations. In other words, the intra-active nature of musicality—its responsive doing of testing and touching—introduces a dis/continuity at the heart of musico-materializations.

A musicality that includes the insensible and unheard challenges our common notion of unmusicality—of failed skills and of failing to master a musical composition according to pre-defined aesthetics. Now, unmusicality is introduced to how a vocalist is encouraged to take the leap into dynamic being, following the lure of a sonority-to-come, and hence the lure of musico-material forces that enrich modes of musicality. Consequently, when not alive to the lure of the unheard, it introduces unmusicality as an indifference—of not caring for sonority-to-come, and thus not enabling the responsiveness of musico-material Others-to-come. As felt and intuitive moments, a vocalist might feel this as not "being fully there" and withholding the desire for sonority. The vocalist may seek solace in *inter*-activity, suggesting control over that which has been rehearsed—a condensed past brought into play, pre-existing the musical event. As a personal account from my own singing practice—a well-known experience of many colleagues—I experienced such indifference of not caring to imagine and not following the lure of sonority-to-come when I premiered new compositions. Although feeling pleased that, after an intense period of rehearsals, everything went well, I knew something was missing; I could not stay with the tone, embracing its full potential; I could not be alert and "sing" with the audience, enabling them to "sing" with me; and I could not be alert and playful enough to imagine musico-material Others as *sym*phonia-to-come.

Yet, when "being there" in the midst of musico-materializations—being alive to an intra-active generated musicality at the edge of musico-materialization that affirms an onto-ethico-aesthetic force from the depths of heterogeneous musico-material doings—means nevertheless that neither unmusicality as indifference nor the musicality of "being there" suggest a horizon, either of the justness or the beauty of music. Rather, being an inseparable part of musico-materializations confronts us with an entangled un/musicality. This involves how, in the words of violinist Isaac Stern, being "the eternal optimist" does not come from the achievement of perfection in a given piece but rather, paradoxically, "from having lived a lifetime in the constant company of the impossibility of duplicating a single performance" (Stern, cited by philosopher Karen Houle 2005: 92).

Although she does not use these exact terms, Houle (2005) affirms an entangled un/musicality as a musico-material normative and differential responsiveness:

> Stern's perpetual attending-to the vital fact of a tentative outcome—the thing never fully given in advance but that must be risked each time and fully—turns out to have been not a source of weariness and discouragement, but the very thing that allowed him perpetually to believe that the next time he performed a piece it might be a little better. His was a "belief" *towards* the better, but not *in* the better.
>
> (92, emphasis in original)

In conclusion, Stern's eternal optimism and Houle's assessment of it, embracing the vocalist's leap, demonstrate that we—performers, listeners, dancers, producers, musicologists, politicians, in short, all (and everything) who are intra-actively called into musico-material being—cannot promise "better" music. We might, however, say that "better" music is part of any musico-materialization-to-come, urging us to be wary of the politics of putting one's trust in repeatability, capturing the "better music" and controlling a future. As demonstrated by the vocalist, past and future are taken up in an intra-actively produced musicality—the onto-ethico-aesthetic force of differential responsiveness that is local, temporal, material, meaningful, *and* speculative and unforeseen. In other words, a vocalist's politics urges us to be alive to the slash of un/musicality as intra-actively integrated past/future condensations, creating new and un/expected temporal and local determinations of all that is involved in *this* music-in-the-making and in what a vocalist *can* (or must not) sing.

With Barad, using my experience as a vocalist, I elaborated the conditions under which the vocalist, vocal body, and musico-materialities could speak through and from within the musico-material act. Being alive to an intra-actively produced un/musicality in all directions implies musico-material experiencing to be more intimate, suggesting "us" to be an inseparable part of, and creatively involved in, the "doing" of musicality. An intra-active musicality challenges the concept of linear time and set borders, orienting our singing, research, and theorizing to start from the relationalities through and within which the musico-material comes to matter. Acknowledging this chapter's limited palette, my hope is nevertheless that thinking with un/musicality, rather than ensuring some resolution to existing problems of musico-materialization, opens to new questions, new kinds of musico-material involvements, and new kinds of musico-material theory and practice encounters, emphasizing a relational onto-ethico-aesthetics of intra-actively holding together.

Notes

1 This chapter builds on my dissertation *Music/ology: A Baradian Account* (Pranger 2020), in which I explore a musicological knowing and valuing in being, reading together the insights of musicologist Carolyn Abbate and physicist and philosopher Karen Barad.

2 The vocal body incorporates the broader musico-material apparatus. When necessary, I define its contours more specifically.

3 Voice studies scholar Nina Sun Eidsheim (2015) discusses this logic of standardization in terms of "[s]onic reductions [being] the tendency to constrain our understanding of sound through previously defined referents" (2).

4 Musicologist Michael Gallope (2017) seemingly makes a similar point when arguing that, although musicians "work outside centralized paths, [they cannot make music] without a historicity, without a form, without an adoption of the past" (256). However, the point I wish to make here concerns the embraced complexity of being caught up in dynamic musico-material doings through and within which an assumed difference between outside and inside becomes (re)configured.

5 Representationalism is based on a two-world logic that assumes a pre-defined distinction between human knowledge, human language, other symbolic means, and an outside world, leading to questions of correspondence.

6 In their introduction to the special edition of *Contemporary Music Review* 39, no. 5 (2020), on "Musical Materialisms," Matthew Sergeant, Isabella van Elferen, and Samuel Wilson (517–525) point to the way in which Barad contributes to musicological scholarship as demonstrated through the articles of Luc Döbereiner (2020: 602–617) on contingency as an inseparable part of musico-material doings; and Lauren Redhead (2020: 564–579), who elaborates Barad's notion of ongoing material liveliness, addressing a voice's material and spatial dimensions.

7 Barad (2003, 2007) elaborates Donna Haraway's (1997) concept of diffraction both as a methodology and as a method.

8 For further reading on diffraction as a concept, method, and methodology, see Kaiser and Thiele (2014), editors of *Parallax 20,* no. 3 (2014), in a special issue on diffraction.

9 Here, I read Barad's (2007) concept "ongoing historicity" (151) in the earlier citation in terms of the concept "historiality" (393). Although historicity refers more to the fact that things change and are infused with their past, with regard to Barad's insights, the notion of "historiality" involves how forces from past and future pass through "things," defining their stability and instability as relative (see also Barad 2007: 438, note 84).

10 Aden Evens (2005) uses the word *suitable*, with which, in my reading, he traverses human intelligibility as a pre-empted condition for materializations. Evens argues that the concert space is not empty. Rather, it is already full of vibrations, waiting "to be modulated to the point of audibility by a *suitable* sound" issued from a vocalist's mouth (Evens 2005: 7, emphasis added).

11 See also Tiainen (2007: 157–158) on the musico-materialization of concepts in a singer's training.

12 The relationship between the philosophy of Barad and Deleuze has increasingly become an object of scholarly attention. See Karin Murris and Vivienne Bozalek (2019: 872–886) for a "response-able reading" of Barad's and Deleuze's insights on a relational ontology.

13 Musicality as a normative and creative force contributes as an intra-actively produced responsiveness to (musico)spacetimematterings, in which "mattering" is understood as the marking of spacetime, matter, music, and meaning. That is to say, time and space are not parameters outside musico-materialization, but are enfolded in a "dynamic field of [musico-] matterings" (Barad 2012b: 221, note 16).

14 The slash disrupts the word dis/continuity performatively, suggesting that matter is "neither fully discontinuous with continuity [n]or even fully continuous with discontinuity, and in any case, surely not one with itself" (Barad 2010: 244). Gamble et al. (2019) caution that the slash might be misread as "invoking a both/and kind of logic ... rather than the neither/nor logic" of Barad's performative account of matter (131, note 121). Crucial here is the manner in which the slash troubles absolute separation: even the cut as a temporal and local materialization of continuity or discontinuity cuts together apart as the separability of differences within (Barad 2012a: 22).

References

Abbate, Carolyn. 2001. *In Search of Opera*. Princeton: Princeton University Press.

Abbate, Carolyn. 2004. Music—Drastic or Gnostic? *Critical Inquiry*. 30 (3): 505–536.

Alaimo, Stacy, and Susan Hekman, eds. 2008. *Material Feminisms*. Bloomington: Indiana University Press.

Barad, Karen. 2003. Posthumanist Performativity: Toward an Understanding of How Matter Comes to Matter. *Signs: Journal of Woman in Culture and Society*. 28 (3): 801–831.

Barad, Karen. 2007. *Meeting the Universe Halfway: Quantum Physics and the Entanglement of Matter and Meaning*. Durham, NC: Duke University Press.

Barad, Karen. 2010. Quantum Entanglements and Hauntological Relations of Inheritance: Dis/Continuities, SpaceTime Enfoldings, and Justice-to-Come. *Derrida Today*. 3 (2): 240–268.

Barad, Karen. 2012a. Intra-active Entanglements—Interview with Karen Barad. *Kvinder, Køn & Forskning*. 21 (1–2): 10–25.

Barad, Karen. 2012b. On Touching: The Inhuman that Therefore I Am. *Differences: A Journal of Feminist Cultural Studies*. 23 (3): 206–223.

Barad, Karen. 2012c. *What Is the Measure of Nothingness? Infinity, Virtuality, Justice*. Kassel: Hatje Cantz.

Barad, Karen. 2015. Transmaterialities: Trans/Matter/Realities and Queer Political Imaginings. *GLQ: A Journal of Lesbian and Gay Studies*. 21 (2–3): 387–422.

Bergson, Henri. [1889] 1910. *Time and Free Will: An Essay on the Immediate Data of Consciousness*. Translated by F. L. Pogson. Open Library: Nabu Public Domain Reprints.

Braidotti, Rosi. 2012. Interview with Rosi Braidotti. In: Rick Dolphijn, and Iris van der Tuin (Eds.). *New Materialism: Interviews and Cartographies* (pp. 19–38). London: Open Humanities Press.

Brett, Philip. 2006. Musicality, Essentialism, and the Closet. In: Philip Brett, Elizabeth Wood, and Gary C. Thomas (Eds.). *Queering the Pitch: The New Gay and Lesbian Musicology* (2nd ed., pp. 9–27). New York: Routledge.

Deleuze, Gilles. [1994] 2004. *Difference and Repetition*. Translated by Paul Patton. New York: Continuum.

Deleuze, Gilles, and Félix Guattari. [1980] 2004. *A Thousand Plateaux: Capitalism and Schizophrenia*. Translated by Brian Massumi. London: Continuum.

Deleuze, Gilles, and Félix Guattari. 1994. *What Is Philosophy?* Translated by Graham Burchell and Hugh Tomlinson. New York: Verso.

Döbereiner, Luc. 2020. Materiality, Contingency and Emergence of Compositional Material. *Contemporary Music Review*. 39 (5): 602–617.

Dolphijn, Rick, and Iris van der Tuin. 2012. *New Materialism: Interviews and Cartographies*. London: Open Humanities Press.

Eidsheim, Nina Sun. 2015. *Sensing Sound: Singing & Listening as Vibrational Practices*. Durham, NC: Duke University Press.

Evens, Aden. 2005. *Sound Ideas: Music, Machines, and Experience*. Minneapolis: University of Minnesota Press.

Gallope, Michael. 2017. *Deep Refrains: Music, Philosophy, and the Ineffable*. Chicago: The University of Chicago Press.

Gamble, Christopher N., Joshua S. Hanan, and Tomas Nail. 2019. What Is New Materialism. *Angelaki*. 24 (6): 111–134.

Haraway, Donna. 1988. Situated Knowledges: The Science Question in Feminism and the Privilege of Partial Perspective. *Feminist Studies*. 14 (3): 575–599.

Haraway, Donna. 1997. *Modest_Witness@Second_Millennium.FemaleMan©_Meets_OncoMouseTM: Feminism and Technoscience*. New York: Routledge.

Honing, Henkjan, Carel ten Cate, Isabelle Peretz, and Sandra E. Trehub. 2015. Without It No Music: Cognition, Biology and Evolution of Musicality. *Philosophical Transactions B. Royal Society*. 370: 1–8.

Houle, Karen. 2005. Micropolitics. In: Charles J. Stivale (Ed.). *Gilles Deleuze: Key Concepts* (pp. 88–98). Montreal: McGill-Queen's University Press.

Kaiser, Birgit Mara, and Kathrin Thiele. 2014. "Diffraction: Onto-Epistemology, Quantum Physics and the Critical Humanities." Introduction to the Special Issue "Diffracted Worlds—Diffractive Readings: Onto-Epistemologies and the Critical Humanities." *Parallax*. 20 (3): 165–167.
Knieter, Gerhard L. 1967. Musicality Is Universal. *Music Educators Journal*. 54 (3): 46–48, 113–116.
Lockheart, Paula. 2003. A History of Early Microphone Singing, 1925–1939: American Mainstream Popular Singing at the Advent of Electronic Microphone Amplification. *Popular Music and Society*. 26 (3): 367–385.
Martinez, F. Sergio, and Alejandro Villanueva. 2018. Musicality as Material Culture. *Adaptive Behavior*. 26 (5): 257–267.
de Mink, Freya. 2013. *Musical Prodigies: Past, Present, and Future Perspectives on Exceptional Performance and Creativity*. Master thesis, Utrecht University.
Murris, Karin, and Vivienne Bozalek. 2019. Diffraction and Response-able Reading of Texts: The Relational Ontologies of Barad and Deleuze. *International Journal of Qualitative Studies in Education*. 37 (7): 872–886.
O'Flynn, John. 2005. Re-appraising Ideas of Musicality in Intercultural Contexts of Music Education. *International Journal of Music Education*. 23 (3): 191–203.
Ojala, Juha. 2014. A Peircean Approach to Musicality. *Interdisciplinary Studies in Musicology*. 14: 175–191.
Panurgiá, Neni. 2018. Anthropism/Immanent Humanism. In: Rosi Braidotti, and Maria Hlavajova (Eds.). *Posthuman Glossary* (pp. 44–45). London: Bloomsbury Academic, Bloomsbury Publishing Plc.
Pranger, Jannie. 2020. *Music/ology: A Baradian Account*. PhD dissertation, Utrecht University.
Redhead, Lauren. 2020. Vibrant Echoes: A Material Semiotics of the Voice in Music by Iris Garrelfs and Marlo Eggplant. *Contemporary Music Review*. 39 (5): 564–579.
Sergeant, Matthew, Isabella van Elferen, and Samuel Wilson. 2020. Introduction to "Musical Materialisms (Plural)". Special issue of *Contemporary Music Review*. 39 (5): 517–525.
Théberge, Paul. 1997. *Any Sound You Can Imagine: Making Music/Consuming Technology*. Hanover: Wesleyan University Press.
Tiainen, Milla. 2007. Corporeal Voices, Sexual Differentiations: New Materialist Perspectives on Music, Singing and Subjectivity. In: Sylvia Mieszkowski, Joy Smith, and Marijke de Valck (Eds.). *Sonic Interventions*. 18 *Thamyris/Intersecting: Place, Sex, and Race* (pp. 147–168). Amsterdam, New York: Rodopi B.V.
Toporova, Alla V., Zarema Sh. Gadzhieva, and Fatima V. Malukhova. 2016. Musicality: A Phenomenon of Nature, Culture, and Identity. *IEJME—Mathematics Education*. 11 (5): 1373–1382.
Trehub, SE, J Becker, and I Morley. 2015. Cross-Cultural Perspective on Music and Musicality. *Philosophical Transactions of the Royal Society B*. 370: 20140096.
Tuin van der, Iris. 2013. Non-Reductive Continental Naturalism in the Contemporary Humanities: Working with Hélène Metzger's Philosophical Reflections. *History of the Human Sciences*. 00 (0): 1–18.

Part II

Voicing intra-active human agencies

3 Trans-corpo-vocality

Voice, gender, and environment in Demian Seesjärvi's transmasculine senses of self

Milla Tiainen

> The ways we constantly affect one another with voice ... are ... really significant, and how all the impulses, the environment, everything affects our experiences of our own voice in turn ... [What matters to me is] how safe I feel and how fully I can be myself, like, it immediately impacts how [I experience my voice] ... [and it] serves as a good navigator for me in social situations in which my voice starts to get more tense ... and I feel that I've lost the connection [to my voice], well then I immediately begin to scrutinize what's going on ... like, is there something in me or something in the other people present that gives rise to my mode of experience.
>
> (Seesjärvi 2020)

The excerpt above is from an interview[1] with Finnish singer and female-to-male (FTM) transman[2] Demian Seesjärvi that was conducted by me (Milla Tiainen) and music and gender studies scholar Taru Leppänen as part of a wider research collaboration among the three of us.[3] In the excerpt, Seesjärvi muses on the relations between his voice, his sense or *senses* of self,[4] and the environment. The last of these terms encompasses fellow humans as well as all sorts of environmental impulses. Seesjärvi first underlines the general power of voice to generate interrelations and attendant effects across individuals and environments as somebody's voicings[5] affect, in one way or another, the other beings within their scope. Reciprocally, a vocalizer's ways of voicing and experiencing themselves may be affected by their relation to a multitude of happenings in their surroundings. Adopting a more personal tone, Seesjärvi then suggests that his vocalizations, or how they sound and feel to him, give him signals about the character of different situations in which he is involved. He maintains that his voice acts as an indicator of features—also negative ones—that inform the unfolding relationship between the environments and occasions in which he participates and his senses of self. As an example of such vocal signals, Seesjärvi mentions the tenseness that, in certain situations, he perceives in the sound of his voicings and, by implication, in his voicing body.

DOI: 10.4324/9781003390992-6

Seesjärvi's reflections highlight crucial aspects of the workings and nature of voice. For starters, he points out how, despite their intimacy, one's voicings, concurrent corporeal feelings, and experiences of oneself are not isolated processes. Instead, vocalizations and the voicer's accompanying sensory and subjective experiences are inextricably relational: they emerge from encounters and impact in-between vocalizations and environment, or voicing bodies and selves and their milieus. In terms of his background, Seesjärvi is a professionally educated singer and vocal pedagogue who has taught singing, and voice coached transgender people for the past decade (Seesjärvi 2020, 2021). These activities have provided him with a well-developed sensitivity regarding the occurring relations between his modes of voicing, self-experience, and particular environments.[6]

Moreover, Seesjärvi points to the relationship between one's voicings and *gendered* senses of self. Although it is not obvious from the above excerpt as such, Seesjärvi's surrounding comments during our interview confirm that his references to feeling safe or unsafe and being able or unable to "be himself" pertain to his subjective and social experiences as a transmasculine person.[7] More specifically, Seesjärvi offers that the extent to which his masculinity seems to be accepted—or not—in the given situation and context affects his voicings all the way down to their physiological and sensory minutiae.[8] This underscores the mutually influencing relationships between voicings, the gendered aspects of the voicer's self, and the things taking place in the environment. At the same time, vocalizations themselves emerge from within intricate contextual, situational, physical, corporeal, and psychical—put differently, socio-material—relationships.

In this chapter, based on my research collaboration with Demian Seesjärvi, I explore *the relationships between voice, the voicer's gendered senses of self, and the environments of their voicings as constantly emergent and co-constitutive.* My understanding of senses of self is similar, for example, to ethnomusicologist and voice researcher Katherine Meizel's (2020: 1–7) and transgender studies scholar Eva Hayward's (2010) theorizations of the multidimensional, ever-emergent relationships between selfhood, (trans)gender, environment, and voice even in the case of "one" person. Thus, with senses of self, I mean a person's accumulating, yet dynamic and multifaceted, experiences, along with ideas of who they are and what they can do, which take shape in a co-constitutive relationship with varying milieus and responses from other beings. As Seesjärvi's accounts below exemplify, senses of self are narratable (cf. Meizel 2020: 7). However, following new materialist theories in feminist research, the study of art, and other fields, I understand them to also include such material changes in a person's relationships, state, and sensory experiences that impact, yet partly overflow, the narrative and discursive means with which senses of self are pinned down in language (see, e.g., Alaimo and Hekman 2008: 2–7; Barrett and Bolt 2013: 5–8; Tiainen 2007: 55–58).

Numerous studies in the arts and humanities have examined from different perspectives the interdependence of vocal renditions, their contexts and attendant notions of selfhood, agency, and gender, along with other identity markers (Cusick 1999; Cavarero 2005; Bosma 2013; Cavarero et al. 2018; Eidsheim 2019; Thomaidis 2020). I approach this topic area via two questions whose exploration can shed new light on the co-constitutive relations of voice, gendered selfhood, and environment. First, in what particular ways may this relationality matter to a transmasculine person? How may the co-constitutive relations of voice, gender, and environment occur in a transman's voicing activities, and how might he experience these relations as either enriching or undermining his gendered senses of self?

Second, I ask what kinds of materialities may participate in voice–gender–environment relations. How can different materialities play a formative and, thus, *agential* role in a transman's voice-informed senses of self? In light of Seesjärvi's reflections above, these materialities include multiple elements and scales, ranging from co-present vocalizing and sensorially experiencing human bodies to the more minute workings of bodily matter and from vocal vibrations and sounds to the social, but also material, happenings in the given environments. Because it is impossible to cover all forms of everyday voicing in one chapter and because singing holds a special place in Demian Seesjärvi's activities and senses of self, I will limit my attention to different *material, socioculturally contingent practices of singing,* as discussed by Seesjärvi.

Toward a new materialist voice and transgender research

My starting premise is that the transmasculine senses of self this chapter explores emerge through the interrelations among (sung) voicings, subjective, and social experiences of gender and material-sociocultural environmental processes. The theoretical underpinnings of how I approach voice and environment are detailed below. A key guideline for understanding gender here is the idea of transmasculinity—or any form of (trans)gender—as diversity. This refers to the significant variation in the relationship of transindividuals to established manifestations and binaries of gender and hormonal and/or surgical means of affirming one's gender (Cromwell 2006; Meizel 2020: 140–141; Prosser 1998). Transgender senses of self also greatly vary depending on transpersons' sociocultural backgrounds, their attachment to intersecting identity markers such as sexuality, class, race, age, and religion, and the activities and communities in which they participate (Stryker 2006: 15; van Midde et al. 2018: 5). For example, Demian Seesjärvi unequivocally identifies as a man instead of being nonbinary or genderfluid. Simultaneously, he considers the psychophysical process of transitioning integral to his manhood. Indeed, Seesjärvi is "out" as a transman because he acknowledges that this form of gender is crucial to his subjective, social, and material experiences of being a man and a voice professional. He has repeatedly discussed his gender in the Finnish media to advance public awareness and acceptance of transpeople. In addition, as studied below, Seesjärvi's long-term relationship with Western classical singing and several other vocal styles has intimately affected his transmasculine senses of self (see also Leppänen and Tiainen 2018).

My approach to gender also draws on the broad but essential starting point proposed by one of the pioneers of interdisciplinary transgender studies, Susan Stryker. According to Stryker, transgender studies scholars do well in mapping the manifestations of gender as taking form via socially normative as well as changing linkages "between the biological specificity of the sexually differentiated human body, the social roles and statuses that a particular form of body is expected to occupy, the subjectively experienced relationship between a gendered sense of self and social expectations of gender-role performance, and the cultural mechanisms that work to sustain or thwart specific configurations of gendered personhood" (2006: 3). This definition underscores the mutually influencing relations of the biological, bodily, social, cultural, and subjective dimensions of gender. Although they vary across and within human groups, these relational dimensions arguably tend to be different for transpersons than for cisgender persons. This is the case, for example, because of the gender dysphoria experienced by many transpeople, Demian Seesjärvi included, and the often-high pressure they feel to "pass" as socially legible members of their gender (Green 2006; Catalano 2015).

While leaning on these notions of gender, I seek to enrich them with new materialist theoretical insights. In the introduction to *The Transgender Studies Reader*, Stryker discusses how conceptions of gender, which have been dominant (especially in Western societies) for the past few centuries, rest on understandings of the material world developed within modern Western-originated scientific materialism. Matter, in these understandings, comprises the world's primary, real stuff that predetermines the types of knowledge that humans can acquire about the world. In the case of gender, this matter that "ultimately matters," according to the views in question, concerns the biological and physiological features of bodies conventionally called "sex" and categorized in terms of the male–female binary. These material features are assumed to precondition the gendered behaviors and characteristics of individuals with a particular kind of body (Stryker 2006: 9).

New materialist theories offer a very different understanding of how materialities matter, including those of human bodies, voices, and (trans)genders. Unlike the scientific materialist views discussed by Stryker, these theories do not consider the material traits of bodies or "sex" as a stable dualistic base that would predetermine the "superstructure" expressions of gender.[9] According to new materialist perspectives, materialities, ranging from bodily matter and processes and vocal emissions to different forms of technology and nonhuman nature, display their own distinctly material ways of acting, reacting, and obtaining features. Yet, these ways are not ontologically self-contained or preordained. Instead, materialities are best understood in terms of open-ended tendencies that come into being via co-constitutive relations with other things: other material entities and social, cultural, and discursive activities (Barad 2007; Coole and Frost 2010; Kontturi 2018: 82–96; Gamble et al. 2019: 122–127).

I bring this idea of properly material yet dynamic and relational materialities into examining how voice–gender–environment relationships shape a transmasculine vocalizer's senses of self. This proposition is encapsulated in the chapter's titular concept of *trans-corpo-vocality*, which I elaborate in my analysis below. Let me note for now that I am devising this concept at the crossroads of Demian Seesjärvi's experiences of singing and my interpretations thereof, as well as theoretical strands from voice studies, new materialisms and transgender studies. Regarding interdisciplinary voice research (Macpherson and Thomaidis 2015; Eidsheim and Meizel 2019), I base the concept of trans-corpo-vocality on insights regarding the voice's *transversal* nature. This qualifier refers to the widely theorized capacity of sounding voices (and other sounds) to traverse and blur, both materially and conceptually, such binary distinctions as individual body/subject and environment, inside and outside, and animate and inanimate. Via this capacity, the voice profoundly unsettles the modern, especially Western, idea of human individuals and their gender or other identity markers as discrete, self-standing units (Neumark 2010; Mazzei and Jackson 2017).

I focus on exploring what this transversal nature of voice might mean to a transmasculine singer's senses of self. To do this, I combine said notion with the concept of *trans-corporeality,* which entails particular understandings of the body and environment. According to environmental humanities and new materialist scholar Stacy Alaimo, who originally coined the concept, trans-corporeality signifies "the time-space where human corporeality, in all its material fleshiness, is inseparable from 'nature' or 'environment'" (2008: 238). Thus, the concept zooms in on the ways in which the material and experiential features of human bodies take shape in an ongoing relation with the traits and events of their environments, from physical particles to sensory stimuli, or via transits between the given environs and bodies, because bodies also affect their milieus, in turn. Alaimo foregrounds the mutual dependence of human and other living bodies and natural environments in the age of human-caused ecological emergency. Without forgetting this critical aspect, in my case, the concept of trans-corporeality helps examine the inseparability of a transperson's voicing body from its—simultaneously material and sociocultural—environments.

Finally, my conceptualization of trans-corpo-vocality is inspired by recent studies of transgender voices that, instead of representing voice therapeutical approaches, delve into transpersons' lived experiences of voicing activities, such as singing, and how these contribute to their personhood. Addressing voices as "material relationality," gender studies scholars Holly Patch and Tomke König (2018) call such lived aspects trans* vocality (see also Krell 2019; Meizel 2020: 137–158; Constansis 2013). My research resonates with Patch and König's attempt to merge the phenomenologically oriented study of human corporeal experiences and their articulation in language with (feminist) new materialist concepts, such as the idea of agencies as the result of intra-action: co-constitutive relations. A slight difference is that where Patch and König (2018: 32–34) stress the role of transpeople's bodies and intersubjective human relations in trans* vocality, I also inquire into how materialities and entities other-than-human—be these hormonal substances, vocal vibrations, sound technologies, or physical and institutional environments—co-produce human experience, such as a transmasculine voicer's senses of self.

To summarize, trans-corpo-vocality embodies a twofold aim that is entangled with the research questions presented above. First, with this concept, I aim to expand scholarly attention to the significance of voicings to transmasculine singers and transgender individuals more broadly. Thus, the concept is an emergent vehicle for practicing "transgender-conscious" and "transgender-empathetic" voice scholarship, to apply the wordings of music scholar Susanna Välimäki (2017: 327). Second, with its beginnings in multiway encounters between situated empirical materials (concerning Demian Seesjärvi's singing practices) and impulses from transgender studies, voice and music studies, and new materialist theorizing, the concept of trans-corpo-vocality can provide new insights into these areas as a result of their intra-action within my research.

My primary research data consists of the outcomes of my and Taru Leppänen's collaboration with Seesjärvi. These outcomes include several semistructured in-depth interviews (duration 90–120 minutes), Seesjärvi's written reflections on his voice and gender and responses to my and Leppänen's analyses, informal online and offline discussions, and Seesjärvi's recorded vocal performances that can be found on his YouTube channel. Additional research data includes Seesjärvi's MA thesis in singing pedagogy concerning transmen's experiences of vocal change and his recent professional writing on a trauma-sensitive singing pedagogical practice he has been developing (Alaimo and Hekman 2008, 2021). I also engage with several of his interviews in the Finnish media concerning his identities as a transman and voice practitioner. For the purposes of this chapter, I will selectively work with examples from these heterogeneous but thematically connected materials.

In terms of research methodology and ethics, I draw on a mutually enriching link between Karen Barad's notion of ethico-onto-epistemological *cuts* and the guidelines proposed by trans studies scholars. By cuts, Barad means the way that any ideational or material apparatus used in research temporarily halts the studied phenomena, which are, ontologically speaking, always in process, into objects of observation. These cuts are onto-epistemological because researchers do not make them outside of, but as implicated in, the happenings they are exploring. Moreover, the cuts are *ethico*-onto-epistemological because each mode of cutting has consequences for how phenomena can—or cannot—be perceived, experienced, and dealt with. This is an ethical matter (Barad 2007: 175–179, 186, 377–381). I aspire to ethico-onto-epistemologically responsible cuts, first by inviting

Seesjärvi to respond—also critically—to my analyses in their different stages. Thereby, I aim for accountability and reciprocal (though never perfectly equal) agency within researcher–researched relationships, which has been advocated as a vital epistemological and research ethical guideline in methodological considerations of transgender studies (Raun 2014).

Second and relatedly, I avoid instrumentalizing transpeople as mere points of access to discoveries about voice, gender, or other topics, of which several trans studies scholars have warned (Namaste 2009; Kondelin 2014). My aim is to acknowledge the rich nuances of Seesjärvi's "experiential truths" (Meizel 2020: 7) instead of exploiting them as mere illustrations of theoretical concepts. Avoiding instrumentalization is crucial because of transpeople's often socially precarious position and because of my situatedness as a cis female music and voice scholar who is committed to being an ally to transpeople but has no personal experience about living as a transperson.

The analysis unfolds in three parts. The first section considers Seesjärvi's experiences of bodily and vocal change linked to the gender-affirming hormone therapy (GAHT) he has undergone in terms of their relationship to Western classical singing, which is one of the key socio-material environments of Seesjärvi's voicings. The second section also explores the co-constitutive relations between Seesjärvi's voice, gendered senses of self and classical singing via some examples of how Seesjärvi has been received in different situations belonging to these vocal practices. I inquire into what Seesjärvi's experiences might reveal about classical singing as a set of milieus for a singing transman. I also discuss how these experiences intersect with some other aspects of Seesjärvi's life. The third section focuses on such vocal music and music-cultural environments that have helped Seesjärvi discover and affirm his senses of self as a transmasculine vocalizer and person.

Transmasculine materialities meet Western classical singing

As evidenced by our joint interviews, Seesjärvi's writings and journalistic articles about him, co-constitutive relations of voice, and senses of self are a key preoccupation for Seesjärvi. He focuses on this theme, for instance, in his professional article about the trauma-sensitive voice pedagogical approach he has been developing in his capacity as a singing teacher. As his starting point, Seesjärvi stresses the voice's remarkable role in how people experience themselves in tandem with the surrounding world. He emphasizes that these experiences involve both the meanings that the vocalizers give to their own voices and the responses to those voices by others in the associated environments. Based on these notions, Seesjärvi characterizes singing voice as a voice of one's identity that sonically enacts different facets of the voicer's experiential world and sense(s) of self (Seesjärvi 2021: 153–154).

Resonating with these views, for which Seesjärvi draws support from approaches in voice and music pedagogical, psychotherapeutic, and trauma research, he has repeatedly recounted how activities of singing have been vital to him personally in relation to his developing gendered senses of self. This has been true ever since Seesjärvi's childhood up until and beyond his entrance into the process of affirming his identity as a man. For example, in an article about Seesjärvi's transmasculinity published on a Finnish lifestyle-themed website, he is quoted as saying, "[s]inging acted as my original voice [and conveyed] the state of my soul. I sang and wept a lot in my youth. It was really cathartic" (Hyväri 2018). Another article that discusses Seesjärvi as a transgender singer describes how singing has served as a kind of therapy for him. Practicing singing enabled Seesjärvi to feel alive in his pretransition body, which he otherwise experienced as alien (Kyllönen 2021).

These examples illuminate how activities of voicing can be a valued strategy of self-identity and survival for transpersons, echoing Katherine Meizel's (2020: 157) interview-based research with transgender singers. Moreover, Seesjärvi's references to feelings of catharsis and bodily vitality (when the physical characteristics of his body were still incompatible with his gender) indicate how vocal music-making may foster the subjectivity and pleasures of transpeople, even "under the strictures and effects of the heteronormative [and cis-normative] culture," as Susanna Välimäki (2017: 327) puts it in her study of transgender metal musician Mina Caputo. Clearly, voicing and singing have afforded life-affirming, perhaps even lifesaving, sensory and subjective experiences for Seesjärvi.

However, the co-emergences of Seesjärvi's voice, gender, and related environments have also repeatedly complicated or undermined his senses of self as a man. In light of our research collaboration, this especially pertains to Seesjärvi's involvement in the practices of Western classical singing. Seesjärvi studied this vocal tradition for many years up until earning a master's degree in music pedagogy, with singing being his major. He has practiced a range of repertoires habitually sung by classically trained singers, spanning Lied, opera arias, numbers from musicals, and folksong adaptations. In the below passage from our interview conducted in 2017, Seesjärvi describes the difficulties he encountered as a classical singer and transman when his voice was transforming as a result of the GAHT he was undergoing at the time. Before the gender affirmation process, Seesjärvi had sung as a mezzo-soprano vis-à-vis the voice type system, also called the Fach system, that is integral to the field of classical singing. The changes in his body caused by GAHT coincided with vocal changes, which, as Seesjärvi explains, gradually made his voice categorizable as a baritone, one of the main male voice and character types in the *Fach* system. The following is how Seesjärvi recalls the difficulties at stake:

> I experienced the [vocal] change as colossal and perhaps even tried to somehow deny it because I had worked so hard to build my previous [mezzo-soprano] voice which then … collapsed completely … it was not a simple transformation from one voice type to another …. I had to learn a completely new way of singing, and also the technical aspects

> of my singing transformed. I noticed that the old ways of singing were no longer feasible. My larynx changed so enormously that it started to affect my breathing in a way that caused a lot of tension in my diaphragm and the surrounding muscles. And, of course, the problems with breathing further increased the tension in my throat (...) At one point I nearly gave up, thinking that I'll never be able to make my voice work again. Occasionally I still feel sad because my voice used to be more versatile, it was more powerful, vital and capable than my present voice. My voice is now different and functional as such, but its range has changed completely. (...) However, it is not the same anymore and hence, it has been an immense process to part with my old voice. It's only during the past two years that I have been able to sing as I used to, and it has been a very empowering experience.
>
> (Seesjärvi 2017)

Seesjärvi poignantly portrays his more difficult experiences regarding the co-constitutive relations between his singing voice, transmasculine selfhood, and the music-cultural practices of classical singing. Notably, specific materialities or material processes contributed to these experiences. Recalling the new materialist notion of agential materiality, Seesjärvi describes how the transformation of his body's tissues and organs *with* the sex hormones of GAHT altered his capacities to vocalize and his experiences of himself as a classical singer and man. Seesjärvi stresses the dramatic nature of these changes, although he had, of course, been aware of such forthcoming transformations before starting the therapy. Put in new materialist terms as outlined by Jane Bennett (2010: viii, 39–43, 49–51), these transformations of, or in, the body and its ways of giving voice followed their own kinds of trajectories and tendencies, which surprised Seesjärvi, (initially) unfolding beyond his control. They made him sensorially experience his body and its voicings in novel, disorienting ways. The transformations urged Seesjärvi to relearn the classical singing techniques he had already deeply absorbed during many years of training.

The difficulties described by Seesjärvi resemble experiences more or less common to many transmen, according to other studies about transpeople's lived relationships with their voices. Thus, Seesjärvi's narrated experiences, while unique as such, exemplify the challenges encountered by many transmasculine persons to whom seeking a (singing) voice that matches their sense of self is central to their gender identification (cf. Meizel 2020: 142). Often, the vocal changes experienced by transmen undergoing GAHT are substantial as the muscles and membranes of their bodies involved in voicing reform, while the size of the vocal folds and the larynx increase. The vocal shift also tends to happen over a time span of several years. Consequently, it may take transmen a long time to adjust—or, as Seesjärvi stresses, corporeally and psychologically rebuild—their vocal practice (Krell 2019: 147; Meizel 2020: 143–147). The physiological responses of transmen to the therapy—the changes in their bodies co-emergent with male sex hormones—differ from each other. Thereby, Seesjärvi's reference to enormous changes may attest to the singular actions and reactions of his body's vocal tissues. At the same time, undergoing gender-affirming hormonal therapy is, in and of itself, a social privilege: not nearly all transpersons have the means to access it. In any case, Seesjärvi's views on the co-constitutive changes in his voicings and senses of self during the hormonally assisted gender affirmation process resonate with the reported experiences of other transmasculine singers. Likewise, his ambivalent feelings about parting with his preaffirmation voice are similar to at least some other transmasculine vocalists' experiences (Krell 2019: 148).

What lends specificity to Seesjärvi's experiences is their relation to the sociomaterial environments of classical singing. Several studies concerning the historical development and contemporary studies of classical singing have highlighted that this tradition rests on exacting voice technical and aesthetic standards (Potter 1998: 47–66; Tiainen 2012: 102–106). Modern classical voice pedagogy draws on detailed (Western) scientific knowledge of human anatomy and physiology. Furthermore, there are fine-grained aesthetic criteria pertaining to classical voice production in general (from breathing to positions of the vocal tract and movement between registers) and the features of high-quality voices in each *Fach* in particular. These criteria are reiterated across various practices of classical singing and music, from voice lessons to public performances and from singing competitions to music criticism and recordings. In addition, classically trained voices are strictly gendered insofar as the *Fach* system divides voices into female and male voices in a binary fashion (Clément 2000; Tiainen 2007). Although there is increasing public discussion about the need to understand classical singing beyond the cisnormative gender binary (Buonaiuto 2022; Levin 2022), the traditional voice type categories still prevail in this music culture.

Given these viewpoints, Seesjärvi's difficulties in embracing his transforming trans male voice may indicate the challenges he has encountered when negotiating the strict vocal and gender norms of classical singing, which is one of the main environments of his voicings. Some of Seesjärvi's reflections above, such as his comment about his more capable preaffirmation voice, point to ideals of technical proficiency and powerful voice projection endemic to classical singing. Thus, it seems that Seesjärvi has adopted these cultural ideals into his intimate, materially embedded ways of producing voice and assessing his skills as a singer. This would be understandable, considering Seesjärvi's studies and investment in classical vocal practice. In sum, Seesjärvi's moving description of his vocal change illustrates how the mutually constitutive relations of voice, gender, and environment shape a transmasculine singer's senses of self or his trans-corpo-vocal selfhood. Individual voicings always occur from within wider milieus of perception and evaluation. Simultaneously, the conventions of those milieus stay alive—or are modified and challenged—in the actual material acts, textures, and relations of bodies and voices.

Encountering and challenging gendered systems of evaluating voices

The following moments from an interview we conducted in 2020 shed further light on how the vocal and gender norms of classical singing have affected Seesjärvi's senses of self. Here, Seesjärvi reminisces about some other singers' reactions to him when he was a member of the famous Savonlinna Opera Festival Choir:

> There was … a terribly strong … [atmosphere] "we are men" [uses a lower, more resonant voice] and "they are women" … and I remember being this sort of an oddball … and … I would get comments from others when I was still searching for my identity a little bit and I was feeling, kind of, really sensitive … For example, once when I was in a group of people [comprised of the other choir members], someone just started commenting on my physical appearance sort of wholly sincerely, saying that "it is indeed difficult to tell which one of the two you are [man or woman]".
>
> (Seesjärvi 2020)

A moment later, Seesjärvi continued his reflections on expectations concerning voice and gender in the field of classical singing:

> There is a kind of unspoken expectation [about manifestations of gender in classical music culture], especially … at singing auditions in Germany where I have auditioned, there is that sort of peculiar … for example, at the time when my [male] voice was not so voluminous yet and it was transforming, I received a lot of comments that [urged me to] appropriate a truly masculine [tone] … whereas I found myself to be [gives a little laugh] sufficiently masculine already [MT: exactly] … then, once a colleague told me that I should bring more sex into my voice … these kinds of instances have made me feel inadequate as a man.
>
> (Seesjärvi 2020)

Both of Seesjärvi's memories foreground the transversality of voice: the way a sounding voice occurs in the relational nexus of the vocalizer's bodily, subjective insides, and socio-material surroundings. The moment Seesjärvi vocalized in the situations above, his vocalizations belonged as much to the others present, their perceptions, and the cultural habits and norms implied by those perceptions as to Seesjärvi's personal experiences of his voicings and self. What is striking here is the apparent predominance of the binary understanding and cis norm of gender in the communities and practices of classical singing. In light of Seesjärvi's accounts, the relations between his voice, body, and the given environments and perceivers had recurrently resulted in these responses by the latter that questioned or even rejected his transmasculine vocality and person instead of welcoming them. Based on Seesjärvi's memories, the gender norms of classical singing culture involve both sonic and visual or multisensory dimensions. This is evident in the first interview passage when Seesjärvi underscores the gender binary he had encountered in the Savonlinna Opera Festival Choir by vocal means: when uttering the words "we are men," which stand in contrast to the category "women," he uses an almost mockingly low and loud masculine-coded tone. In the same passage, he also tells how some singers had considered his bodily appearance incompatible with the expected habitus of male singers—perhaps especially the habitus of those singing in the baritone *Fach*, which is Seesjärvi's post-transition voice type? This was because his appearance was seen as falling between rigid binary gender conventions. The second passage highlights the sonic dimension as Seesjärvi recounts how several official assessors and colleagues have deemed the volume and tone of his voicings insufficiently masculine—for example, in terms of lacking an appropriate "sexual" charge.

Seesjärvi's experiences gesture to what music and voice studies scholar Nina Eidsheim calls culturally derived systems of perceiving and categorizing voices and making evaluative assumptions about the vocalizers' identities (Eidsheim 2019: 4, 12). Clearly, such systems have informed the co-constitutive relations between Seesjärvi's ways of being a singer and man and how these ways have been received by authorities and his peers in various environments of classical singing—from an opera choir to vocal auditions. Specifically, the systems of perception Seesjärvi has encountered seem predicated on the assumption that voicings should be immediately classifiable as either male and masculine or female and feminine. This is accompanied by narrow aesthetic and social criteria as to what counts as proper vocal markers of these genders and from which kinds of bodies such manifestations should or should not arise. Based on the reactions Seesjärvi has received, common modes of categorizing voices and voicers in the culture of classical singing appear to have little room for transmen whose embodied voices differ from those of cis men, as Seesjärvi himself has often underlined (Seesjärvi 2017, 2020; Kyllönen 2021).

To return to the concept of trans-corpo-vocality, the material traits of Seesjärvi's trans male body and baritone voice can be seen as playing a distinctive agential role in the situations recounted by him. On the one hand, these traits have deeply (re)shaped Seesjärvi's voice production, as well as his auto-perceptions (cf. Eidsheim 2019: 13) and feelings regarding his voicings. This is indicated by Seesjärvi's above references to how he felt sensitive about his (vocal) gender identity and was still searching for it. On the other hand, the materialities of Seesjärvi's body and voice, which continued to change in tandem with hormonal therapy at the time of our interview in 2020, made him newly inseparable from and exposed to the vocal and gender norms of the daily milieus of classical singing. Here, I am by no means claiming that Seesjärvi's body and vocalizations would somehow be responsible for the discriminating and even shaming responses from fellow musicians he has had to endure. Rather, the material features—volumes, timbres, visual, and kinetic qualities—of his voice and body have brought something specific and enriching to the sites of his voicing, while calling out the often-harsh prejudices around voices and genders to be found within classical singing. Insofar as "vocality … must be received and interpreted by others in order to impact identity" (Meizel 2020: 11), it seems obvious that the occurring relations between Seesjärvi's voice, gender, and their environments in the field of classical singing have frequently had a discouraging effect on his senses of self.

A few qualifications must be added now. First, not all the responses toward Seesjärvi have questioned the validity of his vocality and masculinity. In the same interview, Seesjärvi reminisced about distinctly different reactions when, for example, someone had complimented his rich, masculine vocal sound (Seesjärvi 2020). Thus, the links made between his voicings and the gendered vocal qualities deemed desirable in classical voices have varied. Second, although Seesjärvi stresses the gender norms of the culture of classical singing and music, he has also noted how they remind him of the ones at work in the conservative Laestadian religious community in which he grew up. Conservative Laestadianism is a pietistic Lutheran revival movement that began in the Sápmi region in the mid-nineteenth century and that is known for its disapproval of such "worldly vices" as alcohol consumption, premarital sex, birth control, television, make-up, and rhythm music. According to Seesjärvi (2020), his childhood environment at home and in the wider Laestadian community was characterized by forced binary gender roles and a narrow understanding of masculinity, with no space for feminine-coded features, while he still sought models for his masculine identity in his father's and male friends' appearances and behaviors. Considering Seesjärvi's comparison between classical music culture and Laestadianism, it is important to note the emergence of his gendered senses of self and the obstacles to them in relation to several, albeit mutually resonating, socio-material environments. This is characteristic of the transgender diversity I discussed earlier: transpersons' experiences of the self and world vary according to differences in their socio-material location, be these related to class, ethnicity, age, or music and religion. Such intersecting specificities need to be acknowledged in the study of transgender lives and selves (cf. Stryker 2006: 12).

Finally, the mutual impact of Seesjärvi's voice, gender, and their environments must be understood as multidirectional. Above, I have provided examples of how assessments by others in the vocalizer's milieu certainly affect him materially and mentally, shaping Seesjärvi's auto-listening of his voice and senses of being encouraged or rejected as a vocal and gendered being (Eidsheim 2019: 13). Yet according to the notion of co-constitutive relations or intra-action (Barad 2007) that is promoted by new materialist theories, the elements that relate with one another in a given situation are all shaped by their relationships. These elements may even undergo significant changes because of the influence of their mutual relations. Following this idea, Seesjärvi's transmasculinity may affect the communities and environments of classical singing, in turn, even though the impact might be local; for example, this can be seen when he speaks about the particularities of transmen's embodied voices and embraces them as part of classical music, whether in the media (Kyllönen 2021) or at his voice pedagogical lectures and workshops. Furthermore, it is also the material specificities of Seesjärvi's trans-corpo-vocal ways of being—the hormonally and culturally shaped substance of his body and his practices of projecting voice into the world—that may change the perceptions and meanings of gender and classical voice for those coming into contact with him. Thereby, the materialities of voicing may contribute to the transformation of socio-material environments, discourses, and their discriminatory aspects (Alaimo and Hekman 2008: 4; Thompson and Tiainen 2017: 384).

Vocal plurality as authentic transmasculinity

Classical music culture triggers feelings of unsafety in Seesjärvi to the extent that he often does not enjoy vocalizing therein. As will be discussed further, he has sought other styles and environments of singing that better foster his transmasculine senses of self. Yet he has also commented on his fondness of various classical music repertoires for voice. For example, singing the songs by the late nineteenth- and twentieth-century Anglo-American composers Charles Ives, Ralph Vaughan Williams and Benjamin Britten feels "really good" to Seesjärvi in terms of voice–body connection and living out his ways of being a man. He believes these experiences hinge on the "touch of popular [music and culture]," tincturing these compositions. However, in the same interview, Seesjärvi remarked, laughing, that "there is a traditional German side" to him, too, when it comes to vocal music to which he feels an affinity. He singled out the compositions of Richard Strauss (1864–1949), who is particularly known as an opera composer, because their textures and "harmonic world" touch him deeply (Seesjärvi 2020). Seesjärvi has also discussed the importance that *Liederkreis*, Op. 39 (1840)—the song cycle composed by Robert Schumann (1810–1856) to the poems of Joseph von Eichendorff—holds for him. In a written reflection he shared with me and our collaborator Taru Leppänen, Seesjärvi describes how one of the cycle's songs, *In der Fremde* (In a Foreign Land), reminds him of instances in his early life when he sang alone amid nature in his home region in Finland and how, when interpreting this Lied, he engages with a range of feelings that his lived life so far invokes in him (Seesjärvi 2018). Another important memory for Seesjärvi is his infatuation with the vocality of the renowned Swedish tenor Jussi Björling (1911–1960), whose manner of singing he sought to emulate since early childhood (Seesjärvi 2018).

There are, then, many musical niches within Western classical singing in which Seesjärvi has found support and inspiration for his trans-corpo-vocal senses of self: experiences of himself as a bodily and voicing person whose identity, both subjective and social, is constituted *in relation* to different activities and environments, including those pertaining to singing. However, while such niches are available in the domain of classical singing and music, the same applies all the more to Seesjärvi's relationship with the rich traditions of popular music. To give just a few examples, one song that Seesjärvi characterized as particularly dear to him in his reflection referenced above was "Smile" composed by Charlie Chaplin and first recorded with lyrics by Nat King Cole in the mid-1950s. A recording of Seesjärvi's performance of "Smile" can be found on his YouTube channel and was performed at his concert *Becoming myself* (2017, "*Omaksi itseksi*" in Finnish), whose numbers Seesjärvi had chosen so that they would portray significant emotional and mental stages during his process of affirming his identity as a man. Seesjärvi ended the concert with his rendition of "Smile" because he wanted to leave the audience with "an experience of a smile that envelops all of life's colors" and also because he believes, following the Zen Buddhist meditation tradition, for instance, that smiling is a notable means of survival and a corporeally based way of calming oneself in the face of difficult feelings (Seesjärvi 2018). To give another example, Seesjärvi has underlined his affinity with black gospel music. He has reminisced how, as a child, he listened to gospel songs and vocalists, such as Mahalia Jackson, on the radio in secret because the Laestadian religious community disapproves of its members' engagement with any kind of rhythm music (Seesjärvi 2020).

The vocal styles and performers important to Seesjärvi that have been discussed so far evidence how daily musical practices, along with the related environments, are a constitutive feature of human agency and selfhood in terms of bodily and sensory experiences, emotions, identity, and social contact, as famously examined by sociologist of music Tia DeNora (2000: 2015), among many others. Clearly, the kinds of vocal music described and performed by Seesjärvi have also played a significant role in his senses of self and meaning-makings as a transmasculine person (cf. Välimäki 2013: 2–3). To echo researchers of trans* vocality, Holly Patch and Tomke König (2018: 47), who draw on Karen Barad's new materialist thinking, Seesjärvi has partially emerged as a voicing and gendered subject from within the material-discursive practices of singing and listening. He has been affected by these practices while affecting the vocal repertoires, in turn, by embodying, sounding, and making sense of them and by living through the feelings and empowering experiences of self they have afforded. One must also acknowledge the active, agential involvement of materialities in Seesjärvi's experiences. Besides his body's movements, parts, and tissues and the vibratory, sonic qualities of his voicings, these materialities include, for example, the technological sound recording and transmission devices and processes, from radio to CDs or other storage formats, without which Seesjärvi could not have accessed the vocalities of Jussi Björling and Mahalia Jackson, or many of the above repertoires, in the first place. As a result, the co-constitutive relations between a voicing body-person and their environments, as indicated by this chapter's concept of trans-corpo-vocality, also involve nonhuman, inorganic entities. Human experiences of voice, self, and gender take shape within tangles of relations between organic and synthetic materialities, bodies, forces, and actors (Thompson and Tiainen 2017: 382).

Although Seesjärvi engages with several musical traditions, it is popular music—understood here according to the common definition as commercially oriented and heavily technology-dependent music aimed at wide audiences—that has particularly suited his voice and senses of self in the making. This is how Seesjärvi reflected on the topic in our interview conducted in 2020:

> I have searched for [repertoires] in the field of pop and jazz, popular music … [and I have] sung those repertoires through which … I feel I can express who I authentically am … In these fields, the spectrum of [vocal] possibility is immense, and what is prioritized, insofar as the singer is not a mere victim of commercial forces, is the voice and the person of the individual … The world of pop and jazz is also much more inclusive [in terms of gender than classical music culture].
>
> (Seesjärvi 2020)

A while later, Seesjärvi made a further observation about his relationship to the spectrum of vocal possibilities:

> Since early childhood, I have been good at, sort of, imitating different voices so that I remember how as a child I would listen to and then mimic [such voices] … In a way, my voice has always conveyed the contradiction within me and the different sides of my person … In that sense, it [my voice] has always been authentic.
>
> (Seesjärvi 2020)

In the first passage, the fulfillment Seesjärvi feels when voicing and listening to pop and jazz repertoires is related to his experience that these fields accommodate numerous possible combinations of voice, body, and gender. Unlike the rather categorical expectations concerning (dualistically) gendered bodies and the sounds they should make, which Seesjärvi has encountered in the field of Western classical singing, he sees the interrelations of body, voice, the voicer's self and the reception of these by others as multifarious and permissive of variation in the genres and attendant practices of popular music. According to Seesjärvi, what often counts in popular music are individual assemblages of voice, body, person, and gender. To use the terms of this chapter, what matters is the open-ended nature of trans-corpo-vocal acts bringing together individuals and environments, whereby specificities of particular bodies, voices, and genders, such as those of transmasculine singers, can expand vocal traditions and understandings of gender. Seesjärvi acknowledges the operation of commercial forces within popular music, one of the manifestations of which is the treatment of distinctive vocalities, or artists capable of singing several vocal styles, as a saleable commodity (cf. Meizel

2020: 5–6). Despite commercial motives, Seesjärvi believes in the vocal gender diversity of popular music. This was demonstrated, for example, by his participation in the auditions of *The Voice of Finland*, the Finnish version of the international reality TV singing competition franchise (Uotinen 2017).

In both passages above, Seesjärvi also associates the authenticity of his *own* voice and gendered senses of self with plurality. Instead of assigning a single singing style or set of vocal characteristics as his selfhood's true habitat, Seesjärvi maintains that his authentic voice and self are not antithetical to, but informed by many connections, influences, and potentialities. This is highlighted by his comments about his readiness and capacity to imitate diverse voices. In Seesjärvi's view, it is his enactment of different, often mimicked, vocalities that has allowed him to convey his inner contradiction—which I take as referring to the tension between his experiences of gender dysphoria and gender affirmation—as well as "the different sides of my person," or his emergent senses of self. To interpret these views with the concept of trans-corpo-vocality, Seesjärvi's transmasculine modes of being and feeling as himself have arisen from productive and transversal, boundary-crossing relations between various singing practices, vocal genders and musical cultures, including their materialities and his body, voicings, and experiences. Perhaps, it is Seesjärvi's multiple modes of voicing that let him explore such ways of being a man that feel right to him, in distinction from cis-normative masculinities, and which value the intersubjective, or constitutively relational, "complexity of self and identity," as Katherine Meizel suggests with her concept of multivocality (2020: 6).

Conclusion

My task in this chapter has been to examine, here in collaborative contact with singer and voice teacher Demian Seesjärvi, how the co-constitutive relations of voice, the voicer's gendered senses of self, and the given environments might manifest themselves and what they might mean to a transmasculine voicer in particular. I also explored another, concurrent, question about the agential role of materialities in the context and situation-bound relationships between bodily voice, gendered self, and environment. My findings were enabled by exchanges, such as interviews, with Seesjärvi over a number of years and by textual and sonic materials concerning his singing activities. These findings shed some new light on the co-emergence of voices, selves, and environments, which has been an important concern for examinations of voice across many fields over the past several decades. First, my analysis proposes that the environments of voicing events are best understood as materially and socially multidimensional phenomena. In the case of Demian Seesjärvi's transmasculine vocality and senses of self, these environments encompass social situations and norm-bound conventions belonging to specific cultural milieus, for example, those of Western classical singing and Laestadian religious communities, as much as sound technologies allowing access to diverse voices, or natural sites like the forests in which Seesjärvi sang as a child.

Second, this chapter has brought forward various and flexible ways in which the relations among voice, voicer, and environment can be constitutive of the voicer's selfhood and agency. As my analytical examples have demonstrated, Seesjärvi's experiences of himself as a singer and man have often been discouraged by responses in the milieus in which he gives voice—this especially applies to the cis-normative practices of classical singing. At the same time, he has found a plethora of empowering relationships with different kinds of vocal music, pursuing these in situations and environments ranging from solo concerts to media appearances and private instances of voice production and imitation. Third, throughout the chapter, I have sought to bring attention to the multiple material elements and processes contributing to voice–self–environment relations as experienced by Seesjärvi. These materialities extend from the hormonally and singing technically transforming features of his body and voice to such more-than-human entities as sound reproduction media. The concept devised for this chapter—trans-corpo-vocality—provides one means to study the relationships at stake while mapping their co-constitutive, also material, character. Trans-corpo-vocality highlights the ontological inseparability of voice, voicing body-subject and environments, as well as the meanings of this inseparability for transmasculine, along with maybe other kinds of transgender vocalizers.

As for my methodological choice of focusing on a single transman's experiences, one of the key limitations may be that Seesjärvi's long-term activities as a singer and voice teacher probably heighten the importance of voice–gender–environment relations in his senses of self, compared with transmen or transpeople who have less of an investment in vocal practices. This must be acknowledged when considering my findings. However, Seesjärvi's own voice pedagogical work with transpeople with various gender identifications (Seesjärvi 2015) and the way this chapter connects his experiences with insights from voice studies, transgender studies, and new materialist theories enhances the relevance of his experiences to discussions about voice, transgender selves, and mattering materialities. Indeed, on a theoretical and methodological level, my exploration has presented intra-actions between interdisciplinary voice studies, research on transgender voices and voicers, and new materialisms. The chapter has shown how these areas can mutually enrich each other while jointly providing a way to advance research and awareness on the rich importance of voicings—their relations and materialities—to transpeople.

Acknowledgments

I wish to express my deep gratitude to Demian Seesjärvi for our inspiring collaboration, which has taught me so much about voice, music, and gender, and for his generous, unwavering encouragement of this collaboration. A warm thank you also to our co-collaborator Taru Leppänen for her crucial contributions to the research at hand. Finally, my thanks go to Anne Tarvainen and the participants of the Gender Studies research seminar at the University of Turku and the Åbo Akademi University for their helpful and perceptive comments on earlier drafts of this chapter.

Notes

1 All the indented citations in this chapter are from the recorded, unpublished interviews that I and Taru Leppänen have conducted with Seesjärvi. The translations of the excerpts from Finnish to English are mine.
2 This is how Seesjärvi defines his gender. While being a man, he considers it important personally and politically to highlight his life course and experiences as a transgender person (Seesjärvi 2017).
3 Ongoing since 2017–2018, our research collaboration has explored the roles of voicing, especially singing and vocal music-making, in Seesjärvi's daily life as a transman. Although written by me, the present chapter is indispensably indebted to this multiperson collaboration.
4 By "senses," I emphasize the processual mobility of self-experiences. Although these experiences involve continuities and stratification, they keep emerging and diversifying throughout the occasions and relations in which a person exists. Even though numerous theories of identity especially within and after poststructuralist thinking emphasize process and multidimensionality, "senses of self" was chosen as the key notion here instead of identity because of the former's inherent plurality. For a further definition of senses of self, see p. 3.
5 In this chapter, terms indicating process, events, and action, such as voicing and vocalizing, are usually preferred to that of voice. These and the related terms "voicer" and "vocalizer" are used interchangeably.
6 Seesjärvi himself (2021: 171) has discussed his acquisition of this kind of sensitivity. See also singing studies scholar Anne Tarvainen's (2012: 131–137) discussion of voice-related proprioceptive body awareness that can be honed through practice.
7 Ethnomusicologist Katherine Meizel (2020: 140–142) provides an informative overview of the many forms of trans identities, including transmasculinities, in relation to voice production and singing.
8 The feelings of un/safeness and dis/comfort mentioned by Seesjärvi can be associated with the notion of "home" which transgender studies scholars have used to refer to several, often entangled, aspects. First, home may refer to transpersons' possibilities—or lack thereof—to feel at ease in their bodies, depending on whether the physical characteristics of these bodies align with their gender identification. Second, home has been utilized to highlight transpeople's right to welcoming, safe milieus, and communities in a social reality still frequently ordered by cis- and heteronormative understandings of gender (Prosser 1998; Vähäpassi 2013; Paceley et al. 2020).
9 The untenability of the model of sex as base and gender as superstructure has been criticized in gender studies since the 1980s.

Research data

Demian Seesjärvi, n.d. https://www.youtube.com/user/demianSmusic. Accessed 25 September 2023.

Seesjärvi, Demian. 2015. *Kohti autenttista ääntä ja itseilmaisua. Transmiesten äänenmuutos psykofyysisenä kokemuksena* [Finding an Authentic Voice and Self-expression: Trans Men's Psychophysiological Experiences of Vocal Change]. Thesis for the degree of Master of Music, Music Pedagogy. Helsinki: Metropolia University of Applied Sciences.

Seesjärvi, Demian. 2017. Interview with Taru Leppänen and Milla Tiainen, 3 February.

Seesjärvi, Demian. 2018. Written reflections on personally significant vocal music and Taru Leppänen and Milla Tiainen's analysis-in-progress, 2 March.

Seesjärvi, Demian. 2020. Interview with Taru Leppänen and Milla Tiainen, 18 June.

Seesjärvi, Demian. 2021. "Opettaja lukkiutuneen äänen äärellä – traumasensitiivisen laulupedagogiikan kehittäminen" [Teacher's Encounters with a Stuck Voice]. In *Laulajan keho ja mieli – psykofyysisyys, vuorovaikutus ja luovat toimintatavat laulunopetuksessa* [Singer's Body and Mind – Psychophysical Dynamics, Interaction, and Creative Approaches in Singing Pedagogy], edited by Marika Kivinen, Hanna Lammi, and Leena Unkari-Virtanen, 151–180. Helsinki: Metropolia University of Applied Sciences. https://urn.fi/URN:ISBN:978-952-328-314-5. Accessed 2 February 2022.

References

Alaimo, Stacy. 2008. "Trans-Corporeal Feminisms and the Ethical Space of Nature." In *Material Feminisms*, edited by Stacy Alaimo and Susan Hekman, 237–264. Bloomington: Indiana University Press.

Alaimo, Stacy, and Susan Hekman. 2008. "Introduction: Emerging Models of Materiality in Feminist Theory." In *Material Feminisms*, edited by Stacy Alaimo and Susan Hekman, 1–19. Bloomington: Indiana University Press.

Barad, Karen. 2007. *Meeting the Universe Halfway: Quantum Physics and the Entanglement of Matter and Meaning*. Durham: Duke University Press.

Barrett, Estelle, and Barbara Bolt. 2013. "Introduction: Toward a 'New Materialism' through the Arts." In *Carnal Knowledge*, edited by Estelle Barrett and Barbara Bolt, 1–13. London: I. B. Tauris.

Bennett, Jane. 2010. *Vibrant Matter: A Political Ecology of Things*. Durham: Duke University Press.

Bosma, Hannah. 2013. *The Electronic Cry: Voice and Gender in Electroacoustic Music*. Amsterdam: University of Amsterdam.

Buonaiuto, Danielle. 2022. "Two Years on T: Transgender Baritone Lucas Bouk." *Classical Singer Magazine*, 19 October 2022. https://www.csmusic.net/content/articles/two-years-on-t/

Catalano, D. Chase J. 2015. "'Trans Enough?' The Pressures Trans Men Negotiate in Higher Education." *Transgender Studies Quarterly* 2 (3): 411–430.

Cavarero, Adriana. 2005. *For More than One Voice: Toward a Philosophy of Vocal Expression*. Stanford: Stanford University Press.

Cavarero, Adriana, Konstantinos Thomaidis, and Ilaria Pinna. 2018. "Toward a Hopeful Plurality of Democracy: An Interview on Vocal Ontology with Adriana Cavarero." *Journal of Interdisciplinary Voice Studies* 3 (1): 81–93.

Clément, Catherine. 2000. "Through Voices, History." In *Siren Songs: Representations of Gender and Sexuality in Opera*, edited by Mary Ann Smart (pp. 17–28). Princeton: Princeton University Press.

Constansis, Alexandros N. 2013. "The Female-to-Male (FTM) Singing Voice and Its Interaction with Queer Theory." *Transposition. Musique et Science Sociales* 3: 1–27.

Coole, Diana, and Samantha Frost. 2010. "Introducing the New Materialisms." In *New Materialism: Ontology, Agency, and Politics*, edited by Diana Coole and Samantha Frost, 1–43. Durham: Duke University Press.
Cromwell, Jason. 2006. "Queering the Binaries: Transsituated Bodies, Identities, and Sexualities." In *The Transgender Studies Reader*, edited by Susan Stryker and Stephen Whittle, 509–520. New York: Routledge.
Cusick, Suzanne. 1999. "On Musical Performances of Gender and Sex." In *Audible Traces: Gender, Identity, and Music*, edited by Elaine Barkin, Lydia Hamessley and Benjamin Boretz, 25–48. Zürich: Carciofoli Verlagshaus.
DeNora, Tia. 2000. *Music in Everyday Life*. Cambridge: Cambridge University Press.
Eidsheim, Nina Sun. 2019. *The Race of Sound: Listening, Timbre, and Vocality in African American Music*. Durham: Duke University Press.
Eidsheim, Nina Sun, and Katherine Meizel (Eds.). 2019. *The Oxford Handbook of Voice Studies*. Oxford: Oxford University Press.
Gamble, Christopher N., Joshua S. Hanan, and Thomas Nail. 2019. "What Is New Materialism?" *Angelaki: Journal of the Theoretical Humanities* 24 (6): 111–134.
Green, Jamison. 2006. "Look! No, Don't! The Visibility Dilemma for Transsexual Men." In *The Transgender Studies Reader*, edited by Susan Stryker and Stephen Whittle, 499–508. New York: Routledge.
Hayward, Eva. 2010. "Spider City Sex." *Women & Performance*, 20(3): 225–251.
Hyväri, Hanna. 2018. "Korjasin sukupuoltani ja löysin onnen." *Terve.fi*, 11 April. https://www.terve.fi/artikkelit/korjasin-sukupuoltani-ja-loysin-onnen
Kondelin, Sade. 2014. "Dis/Orientations of Gender and Sexuality in Transgender Embodiment." *SQS* 7 (1–2): 32–43.
Kontturi, Katve-Kaisa. 2018. *Ways of Following: Art, Materiality, Collaboration*. London: Open Humanities Press. http://openhumanitiespress.org/books/download/Kontturi_2018_Ways-of-Following.pdf
Krell, Elias. 2019. "Trans/forming White Noise: Gender, Race, and Disability in the Music of Joe Stevens." In *The Oxford Handbook of Voice Studies*, edited by Nina Sun Eidsheim and Katherine Meizel, 143–164. Oxford: Oxford University Press.
Kyllönen, Tero. 2021. "Oopperalavoille tähdännyt Demian Seesjärvi määritettiin syntymässä tytöksi, mutta vasta sukupuolenkorjaus antoi hänelle oman äänen." *Yle*, 9 November. https://yle.fi/aihe/a/20-10001582
Leppänen, Taru, and Milla Tiainen, in collaboration with Demian Seesjärvi. 2018. "Trans-Becomings in Western 'Classical' Singing: An Intra-Active Approach." *Ruukku – Studies in Artistic Research* 9. https://www.researchcatalogue.net/view/372940/372941 Accessed 7 March 2022.
Levin, Annie. 2022. "Baritone Lucia Lucas Becomes First Openly Trans Singer to Perform at the Met Opera." *Observer*, 21 March 2022.
Macpherson, Ben., and Konstantinos Thomaidis (Eds.). 2015. *Voice Studies: Critical Approaches to Process, Performance and Experience*. New York: Routledge.
Mazzei, Lisa A., and Alecia Y. Jackson. 2017. "Voice in the Agentic Assemblage." *Educational Philosophy and Theory* 49 (11): 1090–1098.
Meizel, Katherine. 2020. *Multivocality: Singing on the Borders of Identity*. Oxford: Oxford University Press.
Namaste, Viviane. 2009. "Undoing Theory: The 'Transgender Question' and the Epistemic Violence of Anglo-American Feminist Theory." *Hypatia* 24 (3): 11–32.
Neumark, Norie. 2010. "Introduction: The Paradox of Voice." In *Voice: Vocal Aesthetics in Digital Arts and Media*, edited by Norie Neumark, Ross Gibson, and Teho van Leeuwen, xi–xxxiii. Cambridge, MA: The MIT Press.
Patch, Holly, and Tomke König. 2018. "Trans* Vocality: Lived Experience, Singing Bodies, and Joyful Politics." *Freiburger Zeitschrift für GeschlechterStudien* 24: 31–53.

Potter, John. 1998. *Vocal Authority: Singing Style and Ideology*. Cambridge: Cambridge University Press.
Prosser, Jay. 1998. *Second Skins: The Body Narratives of Transsexuality*. New York: Columbia University Press.
Raun, Tobias. 2014. "Trans as Contested Intelligibility: Interrogating How to Conduct Trans Analysis with Respectful Curiosity." *Lambda Nordica* 1: 13–37.
Stryker, Susan. 2006. "(De)Subjugated Knowledges: An Introduction to Transgender Studies." In *The Transgender Studies Reader*, edited by Susan Stryker and Stephen Whittle, 1–18. New York: Routledge.
Tarvainen, Anne. 2012. *Laulajan ääni ja ilmaisu* [Singer's Voice and Expression]. Tampere: Tampere University Press.
Thomaidis, Konstantinos. 2020. "Dramaturing the I Voicer in *A Voice Is. A Voice Has. A Voice Does:* Methodologies of Autobiophony." *Journal of Interdisciplinary Voice Studies* 5 (1): 81–106.
Thompson, Marie, and Milla Tiainen. 2017. "Laughing Matter." In *Gender: Laughter*, edited by Bettina Papenburg, 377–391. Farmington Hills, MI: Macmillan Reference USA.
Tiainen, Milla. 2007. "Corporeal Voices, Sexual Differentiations: New Materialist Perspectives on Music, Singing, and Subjectivity." In *Sonic Interventions*, edited by Sylvia Mieszkowski, Joy Smith, and Marijke de Valck, 147–168. Amsterdam: Rodopi.
Tiainen, Milla 2012. *Becoming-Singer: Cartographies of Singing, Music-Making and Opera*. PhD dissertation, University of Turku.
Uotinen, Jenni. 2017. "Sukupuolensa korjannut Demian kisaa tänään The Voice of Finlandissa: Olen kärsinyt häpeästä ja pelosta." *Iltalehti,* 16 February. https://www.iltalehti.fi/tv-ja-leffat/a/201702142200070496
van Midde, Max, Vick Virtù Ludovico, and Olga Cielemęcka. 2018. "Editorial – Trans Materialities." *Graduate Journal of Social Science* 14 (2): 4–9.
Välimäki, Susanna. 2013. "Listening to Transgender Utopia in Boys Don't Cry." *SQS* 1–2: 3–17.
Välimäki, Susanna. 2017. "Confronting the Gender Trouble for Real: Mina Caputo, Metal Truth and Transgender Power." In *The Routledge Research Companion to Popular Music and Gender*, edited by Stan Hawkins, 326–346. New York: Routledge.

4 "The body vibrates with the power of singing"

Material, somaesthetic, and transactional features in the vocal experiences of singers with acid reflux

Anne Tarvainen

In singing, the singer's body and surroundings provide the material basis for the activity. Through practice, the singer learns to sense subtle nuances of their material reality and use this feedback to regulate their vocal output. These materials, including the tissues of the singer's body and the surfaces of their environment, not only shape the singer's perception but also emerge as potential actors in the singing process. In this chapter, I will examine how materiality manifests in different vocal experiences of professional and amateur singers with acid reflux by analyzing their written responses to an internet survey conducted as part of my research project, "The Feel of Singing," from 2018 to 2021.[1] Singers with vocal problems have traditionally been studied through the frameworks of pathology and disability, focusing on general patterns within large populations. My aim is, however, to build on the singers' unique perceptions of materiality by exploring the issue from perspectives beyond physiology and focusing on the experiential features that indicate either vocal problems or "successful singing."

Laryngeal reflux[2] is a common cause of vocal issues, and singing can predispose individuals to reflux due to the increased abdominal pressure required by most vocal techniques (Koufman et al. 2000; Lechien et al. 2019: S41; Lenti et al. 2019: 798, 802; Merati 2020: 196). Given that the respondents in this study suffer from acid reflux, many also experience vocal problems that hinder their singing. However, many of these singers also report moments where singing is enjoyable and meaningful. Setting aside prior assumptions about these singers' perceptions or abilities, I will explore the spectrum of their unique vocal experiences, revealing both the profound value of singing in their lives and the intricate challenges posed by their vocal issues.

The topic of embodied vocal experience and its materiality will be viewed through the lens of *vocal somaesthetics* (Tarvainen 2016, 2018a, 2018b), which draws on *pragmatist somaesthetics* (Shusterman 2008, 2012b), *interdisciplinary voice studies* (Thomaidis and Macpherson 2015; Macpherson 2021), and *new materialist approaches to singing* (Fast 2008, 2016; Tiainen

DOI: 10.4324/9781003390992-7

2012, 2013, 2017; Laasonen Belgrano 2018, 2021; Leppänen 2018; Leppänen and Tiainen 2018; Magnat 2019). At first glance, somaesthetics and new materialism may appear to be distinct approaches, with the former emphasizing embodied experience and the latter focusing on materiality. Despite their differences, these approaches share some common ground, making integrating them in the same investigation possible.[3] The link between experience and materiality becomes accessible through the pragmatist philosophy underlying somaesthetics, offering insights into aesthetic experience and transactionality (Dewey and Bentley 1949; Dewey 2005/1934). In the first two sections of this chapter, I will provide a theoretical foundation for my analysis, and in the following three sections, I will examine the problematic and the best vocal experiences of singers with acid reflux from the perspective of materiality.

Materiality and somaesthetic experience

As mentioned above, somaesthetics and new materialism place differing emphases on embodied experience and materiality. While somaesthetics focuses on the human embodied experience, new materialism, particularly as it relates to post-humanism, explicitly seeks to move away from a human-centered perspective. In this framework, the human body and its materiality are seen as equal to other materials of the world (see, e.g., Ferrando 2012), with the sensory nature of matter often considered beyond the human body (see, e.g., Dolphijn and van der Tuin 2012: 59). Typically, human experience has not been the primary focus of new materialist studies (Lennon 2019).

Although somaesthetics centers its primary focus on human embodied experience, it emphasizes a relational perspective that considers the material body as intertwined with its environment and culture (Sullivan 2001: xvi, 51, 53; Shusterman 2011: 314, 2012b: 227). For Shusterman, the body is a palpable and perceptible entity comprising solid mass, liquids, and gases (Shusterman 2012a: 7). He describes how the body is deeply intertwined with its environment. It fulfills its basic needs by interacting with elements of the environment, such as breathing air and nourishing itself with food. These environmental substances flow in and out of the body, with nutrients becoming an integral part of it (Shusterman 2008: 214). In other words, humans consist of the same matter as other organisms, a notion rooted in Dewey's soft naturalism, differing only in how matter is organized and the complexity of its organization (Shusterman 1994: 129).

Even the human mind is fundamentally material, composed of the same elements as the rest of the world, thus being part of nature (Määttänen 2017: 93). This inseparability of matter and mind is inherent in Shusterman's concept of the *soma*, defined as "the living, sentient, purposive body [...] as the indispensable medium for all perception" (Shusterman 2012b: 3, 227; see also Shusterman 2008: 1–2).

If the soma is entangled with its surroundings, so is the somaesthetic experience, which can be transmaterial, extending through the material boundaries, including those of the human body. Experience involves not only one's body but also the surrounding material environment. As we perceive our bodies, we also feel, at the very least, the air we breathe, gravity, and any surfaces in contact with the body (Shusterman 2008: 8, 98). Beyond the body's materials and environment, experience is also inseparable from action. Sensation and action operate in tandem within experience (Määttänen 2015a: 87, 2015b: 21). The environment is not a passive entity either, as it actively responds and thus shapes the organism's actions and experiences. Moreover, organisms possess physiological plasticity, with actions continuously reshaping their physiology (Alhanen 2013: 60, 66).

From a new materialist perspective, experience is, in a similar manner, material, dynamic, relational, and co-constituted with the world. This view challenges the notion of experience as purely internal, subjective, or even exclusively human. Instead, it emerges through ongoing intra-actions and mutual entanglements of human and non-human entities, sensory perceptions, affective responses, material-discursive environments, and socio-culturally defined situations. Tamsin Jones, discussing religious experience from a new materialist standpoint, argues that experience is "never purely interior and private, for it is writ in our actions, our orientations, and turnings, marked on our bodies and our expulsions, generative of energies and social movements that track one way or another" (Jones 2016: 18). In Karen Barad's agential realism, agency is not an inherent ability of humans or non-humans but rather an enactment that can involve either. It can be understood as a form of reactivity or sensitivity—"response-ability" (Suzuki-Bosco 2016). This suggests that agency is not a self-contained force driven by control but involves openness, receptivity, and influence. In Dewey's philosophy, experience is defined as an interplay of doing and undergoing (Dewey 2005/1934: 45–58), with "undergoing" resonating with Barad's "response-ability."

Action and matter are inseparable in aesthetic experience and expression as Dewey describes them. The materials of the world can become art when transformed by a living organism's energies. This synthesis of matter and action is defined by Dewey's concept of the *medium*—not to be understood as any material channel independent from the action. It arises from the material and the artist's actions applied to it, as Thomas Leddy points out (Leddy 2019: 45).

Dewey also views the artist as a material being. For example, in dance, the material body becomes a medium of expression through movement. Spontaneous embodied reactions alone do not suffice; the body and movement must work in conjunction to transform the body's state and allow the medium to emerge (Dewey 2005/1934: 66–67, 209, 290–291, 299). This change can also occur in singing, where the singer's material body and actions transform beyond everyday movements and vocalizations. Similarly, new materialism views the aesthetic experience as arising from relations between things—whether human or non-human—rather than being located solely within objects or experiences (Bolt 2013: 6). Applied to singing, this relational perspective diverges from conventional Western classical and popular music traditions, which often focus on the voice as an aesthetic object.

The *somaesthetic experience* is both embodied and aesthetic, though not all embodied or *somatic experiences* qualify as aesthetic. Consequently, what constitutes an *aesthetic experience* within this framework? In Dewey's philosophy, aesthetic experience is described as memorable, satisfying, fulfilling, integrated, intensified, heightened, active, dynamic, and immediate. It embodies qualities such as harmony, unity, cumulation, tension, conservation, anticipation, development, and completion. Conversely, *non-aesthetic experiences* may be drifting, yielding, slack, discursive, rigid, dissipated, mundane, loose-ended, coerced, compromised, or incoherent (Shusterman 2000: 27, 55–56; Dewey 2005/1934: 41–42, 48, 51, 64; Ryder 2014: 69; Tarvainen 2018a: 104). However, as an empirical researcher, one need not define aesthetic experience beforehand, as its nature is tied to the material-cultural activities and transactions from which it emerges, taking different forms in different contexts and situations.

With the perspectives of somaesthetics, pragmatism, and new materialism presented here, we can understand the vocal body as thoroughly material, alive, sentient, and entangled with the world—a condition often evident in the aesthetic experiences of singing. Rather than viewing the body as a predetermined unit, we can explore how the materials "belonging to the body" act together to form "internal functional units" or, in new materialist terms, *swarms* (Bennett 2010), which may or may not work in conjunction with each other during vocal acts. Vocal experience is not confined solely to the body but is relationally co-constituted with vocal actions, situations, and environments. This shifts our attention to the diverse material, dynamic, social, and cultural transactions the vocal body and experience stem from.

Singing as transaction

If the actions and experiences of the material human organism are intertwined with the world and culture it inhabits, so is singing inseparable from the singer's material and cultural surroundings. The smoothness, roughness, or swelling of the singer's internal body tissues, the felt lightness or heaviness of singing, the brightness or raspiness of the voice, and its genre-related suitability do not merely exist in relation to each other but are instead inextricably formed together. In other words, the qualitative and aesthetic aspects of vocal sound, as well as the bodily and material aspects of singing, do not exist independently of each other. Of course, vocal problems related to reflux can be viewed causally: a certain kind of food causes a particular type of reaction in the body, which affects the voice in a certain way and makes it difficult to sing. However, in this exploration, I aim to avoid reducing the complex phenomenon of ability, disability, and agency to a one-dimensional cause-and-effect continuum limited to the inner workings of the physiological body. This is where the concept of *transaction* comes in.

The co-constitution of an organism and its environment can be viewed as a *transaction*, as Dewey proposed in his late work with Arthur F. Bentley (1949).[4] While *self-action* refers to the organism's autonomous capacity to act and *interaction* refers to the organism's causal relationship with other beings, transaction highlights the fact that an organism takes shape together with its environment and other organisms. This *co-constitution* is so fundamental that any separate entities, essences, realities, relations, or elements cannot be said to exist before transactions (Dewey and Bentley 1949: 108; Sullivan 2001: 16).[5] Moreover, an organism is never complete but constantly takes shape through various transactions with other organisms and the environment (Sullivan 2001: 1). In this light, the organism is viewed not so much as an "organism-as-whole" but as an "organism-in-environment-as-a-whole" (Dewey and Bentley 1949: 109). Dewey and Bentley have also pointed out that the word "actor" does not capture the complex transactions in which the organism is engaged. They propose that the "actor" should, therefore, always be understood as a "trans-actor" (Dewey and Bentley 1949: 290).[6] Along the same lines, Mark Rimmer explores music, listener, and environment, describing their co-constitution as "music-and-organisms-in-environment-as-a-whole" (Rimmer 2020: 2796). He proposes transactional musicology that could investigate the dynamic configurations of different musical actors—the various flowing and overlapping relationships between interdependent human and non-human agents (Rimmer 2020: 3304, 3313).

In the act of singing, the material body and vocal sounds emanating from it cannot be separated as they emerge together, forming each other. Moreover, the transactions of singing are not limited to the body and voice, as the surrounding space and its various materials, objects, bodies, surfaces, and textures also take shape and mold each other in these mutual processes. In transactions, numerous phenomena occur simultaneously, and their co-constituting stimuli can be manifold. Due to their simultaneity, continuity, and complexity, these co-constitutions are quite challenging to describe in detail. For example, the ensemble we recognize as "a singer" emerges together with tissues, cells, minerals, movements, vibrations, sounds, sensations, vitalities, affects, and countless other factors. At the same time, cells, for example, emerge in response to these same movements, vibrations, minerals, and other factors. Almost all these things arise as they arise, depending on each other's activity and appearance, shaping the overall entity we call singing. The interconnectedness of singing with various materials, movements, objectives, and agencies has been addressed previously through the new materialist approach. Elisabeth Laasonen Belgrano illustrates the *intra-action* of singing, in which different materials and activities are co-constituted. She writes: "[...] no single note stands alone, but always emerges through an intra-active encounter between the vocal apparatus, the body/mind constituency of the singer, the walls, the physical space [...], and of course the various movements occurring continuously" (Laasonen Belgrano 2018).

Looking at vocal problems, we can see that they are not confined to the singer's individual body. Instead, they emerge within numerous transactions where the body becomes—or fails to become—a vocal body. On a personal level, these problems impact not only the singer's physiological performance but also the singer's capacity to experience and connect with both musical textures and fellow musicians, as Milla Tiainen highlights (Tiainen 2012: 88). Tiainen has described how, in order to produce a certain kind of vocal sound in the Western classical tradition, the singer's body parts must work together in a highly specialized way. The body needs to be organized both internally and in terms of musical textures, acoustic space, the air one breathes, aesthetic demands, and audience reactions (Tiainen 2012: 102). The materiality of singing is thus linked not only to the fleshiness of the singer's body and the physical objects of their surroundings but also to the socio-cultural and aesthetic factors involved in the act of singing.

As Shannon Sullivan points out, we are transactionally shaped as part of broader social, political, and cultural factors that affect our bodies (2001: 110–121, 158–159). This does not mean ignoring situationality, which is essential to studying materiality. Following in the transactional footsteps of Dewey and Bentley, we can understand the aesthetic ideals and norms of singing as recurrent, widespread, and slowly transforming chains of actual material transactions. Cultural customs and habits manifest in the singer's flesh through learned patterns of movement, sound, and dynamic environmental entanglements. Thus, the present act of singing is shaped by prior transactions of other materials and bodies, extending historically beyond a given moment. By maintaining a situational inquiry, we can understand how singing becomes singing within a particular material and cultural setting (see also Rimmer 2020: 3362). Similarly, somaesthetics encourages us to avoid outlining general rules about the body or experience, instead examining how they manifest in different activities and cultural settings (Shusterman 2011: 317–318). Such an open approach allows for consideration of the embodied experience without fixed presumptions of what the body inherently is, including its parts, borders, and abilities.

Resisting matter

The following sections are grounded in my research project, "The Feel of Singing," which included an online survey, "Singing and Reflux," with 107 respondents. The participants were predominantly female (83%) and included both amateur (53%) and professional singers (25%), with most aged 35–44 or 25–34 (39% and 22%, respectively). The respondents sang various genres, notably pop (54%), classical (48%), and folk or world music (39%). A significant proportion experienced either gastroesophageal reflux (37%) or laryngopharyngeal reflux (34%), with some having only the latter (21%).

Many felt reflux affected their voice to some degree (58%), while others reported a significant impact (32%). I used multiple-choice questions for demographic and background data, while open-ended questions explored their embodied experiences of singing, its personal meaning, the effects of reflux, and the related feelings, emotions, and thoughts. The aim was not to quantify but to evoke a wide spectrum of experiences related to singing and reflux, analyzed through thematic and transactional[7] approaches focusing on materiality and agency in embodied experiences of singing. My own experiences of singing and reflux also inform this analysis, consistent with somaesthetic research methodologies (Shusterman 2008, 2012b; Tarvainen 2023a).

Reflux symptoms can alter body tissues' texture, thickness, viscosity, moisture, and mucus secretion (Lechien et al. 2019: S40–S41). This can change the vibration characteristics of the tissues and, thus, the vocal sound. The vibrational capacity of the vocal cords depends on their delicate multilayered structure and biomechanical features. However, the mucous membranes of the larynx are easily irritated by components such as acid, pepsin, trypsin, and bile salts (Lechien et al. 2019: S40–S41; Lipan et al. 2006: 261–263). The survey responses revealed that various materials, such as inflamed body tissues, swollen vocal cords, acid, or excess mucus, emerge as actors disrupting the singers' agency. For example, excess mucus or swollen tissue may disrupt singing, introducing unwanted noises and blocking the aesthetic objectives. One singer shared:

> There's a crackling sound caused by mucus. It feels as if acid is rising in the throat from the esophagus.
>
> (ID78–I)[8]

The challenges to agency manifested as rising, leaking, breaking, getting stuck, and other impediments, often during transitions or prolonged actions such as register changes, melodic patterns, or maintaining a pitch. These obstacles highlight how diverse material transactors could conflict rather than align within the singer's agency. Reflux often creates a sense of bodily or vocal resistance. For example, one respondent described their experience as follows:

> The voice cracks and rasps, throat feels tight.
>
> (ID17–I)

Singers provided detailed sensory descriptions of tightness, stiffness, burning, and pain. Interestingly, descriptions often shifted the focus from the singer's personal experience to that of the body itself, as in "the throat feels," reflecting how body tissue can sense its own material qualities.

Compared to other vocal issues, reflux brings more discomfort in the vocal tract (Lopes et al. 2015: 321). The data revealed that reflux causes sensations of irritation, dryness, roughness, itchiness, inflexibility, stuckness, heaviness, swelling, and thickness in the throat, often leading to discomfort, pain, and even agony. The thickened tissues not only bring a sense of constriction inside the body but also reduce the space for resonance, resulting in less controllable sound and more mechanical singing, as illustrated by this respondent:

> With reflux, the voice feels stiff and difficult to control. The voice seems to get 'stuck,' especially at register changes. The vocal cords do not flex, which slows down the singing of the [melodic] patterns and makes it difficult to maintain pitch. Singing becomes mechanical as the basic technique requires so much work. Sometimes, the voice also leaks, and the vocal cords feel swollen. The vocal range narrows. The voice starts to crackle and break. Singing doesn't feel fun; it feels painful and desperate.
>
> (ID68–I)[9]

The issues of agency extend to achieving certain aesthetic goals in singing. The pragmatist approach emphasizes the inseparability of action and sensation in shaping experiences, as perceived things take on meaning according to the activities and goals they are related to (Määttänen 2015a: 87; 2015b: 21; 2017: 92). It can thus be stated that the goal for ideal vocal sounds, the required action, and the embodied material sensations co-constitute in singing. An unexpected challenge in one of these aspects can, at worst, prevent singing. For example, while distortion may be a desirable aesthetic quality of voice in certain vocal genres, it is not necessarily so when it comes to "unhealthy-sounding distortion" caused by reflux, as one respondent explained.

> There's also an unhealthy-sounding distortion that doesn't come with the normal flu and that I can't produce even if I wanted to, even though I have a few different techniques if I want to produce distortion deliberately. At worst, singing during reflux is impossible, so I've given up on it.
>
> (ID107–I)

Singing and distortion can coexist when deliberate but can be problematic when reflux-induced. However, if the aesthetic ideals of a given genre allow it, even reflux-induced roughness of the voice can be acceptable, enabling the formation of an aesthetic experience as well. An example of this is a death metal singer who felt that reflux did not necessarily hamper his vocalization but, on the contrary, brought a genre-appropriate "disfigured" quality to his voice (ID54-2-H). It seems that aesthetic vocal ideals and somaesthetic experiences are intertwined in the transactions of singing. An aesthetically desirable voice can help the singer achieve an aesthetic experience, although the co-constitution of the two may not always be this straightforward.

As an aesthetic activity, singing usually demands precise control of body movements and tissue vibrations, often magnifying reflux-related inflexibilities. The following example shows how the changes imposed by reflux threaten this singer's agency in the transactions of singing but not so much in speaking:

> It's a huge pity when your throat hurts almost as soon as you start singing. As a teacher, I talk a lot, and my voice can take it, but if I sing any louder than with a soft hum, my throat feels miserable.
>
> (ID35–I)

Previous studies indicate that reflux-induced vocal issues may leave speaking voices relatively intact while causing major problems with singing. Singing is more demanding than other vocal activities, and even minor changes in the vocal cords can cause significant vocal problems (Lechien et al. 2019: S41; Nacci et al. 2020: 1). Tiainen describes the carnality of singing, its fundamental materiality, and the control of materials it requires. In singing, the sound is created in the palpable encounter between air and flesh. The subtle control of these moving materials results in pitches, volumes, and other aspects essential to the aesthetics of singing (Tiainen 2012: 104).

The emotional toll of reflux-impacted singing emerged strongly among respondents, expressed through frustration, fear, and anger. The following narrative of a singer highlights a range of negative emotions associated with vocal problems:

> The vocal range is often narrowed momentarily with severe reflux symptoms and only recovers when the throat heals, usually slowly. This causes frustration in me, a sense of failure, and irritation. I also fear that I will not be able to practice my profession to the full or that I will destroy my voice through my own indifference. That is to say, guilt and shame, too, on top of all the above.
>
> (ID145–I)

Ultimately, singers' experiences reflect how the excessive resistance emerging with reflux disrupts artistic intent and aesthetic cohesion. According to Dewey, artistic form is created when the energies of an organism are aligned with the resistances and furtherances of the environment (Dewey 2005/1934: 153). Excessive resistance can cause the disintegration of an experience into seemingly separate parts. In such experiences, stimulus, idea, and response may appear disconnected from each other. A person may experience themselves as an object, separate from the event and its associated agencies (Brinkmann 2011: 306). As Dewey has articulated, the resistance to be overcome in singing is partly within the organism itself. Such resistance may include embarrassment, fear, awkwardness, self-consciousness, or lack of vitality (Dewey 2005/1934: 164–165). In light of the singers' survey answers, it seems that the movements and vibrations within the body, when behaving unexpectedly, create fragmented experiences, leaving singers feeling objectified and disconnected from their voices. In such experiences, the agency is located in specific parts of the body, leaving the singer without the experience of possessing a full and pervasive agency. This can bring difficult emotions to the surface, further fueling inner resistance.

The experiences of vocal problems among the studied singers were notably similar, centering on shifts in voice quality, physiological sensations, and emotions such as frustration, anger, and fear. Most singers articulated these challenges using a physiological vocabulary familiar to vocal pedagogues and methods, highlighting an acute awareness of some parts of their bodies. Agency emerged as a central issue, with tissue changes and excess or misplaced materials from reflux resisting their control and undermining their overall agency. Beyond reflux-induced changes, factors such as the singer's own body, voice, technique, vocal range, and severe reflux symptoms played agential roles,[10] often displacing singers from the center of their experience. The singers lost not only their full agency but also their position at the center of their experience as the sensations perceived by individual body parts and tissues took center stage. Aesthetic goals, intertwined with physical constraints, shaped singers' experiences in complex ways, revealing the situational nature of sound production as the materials, vibrations, intentions, and ideals transacted together. This embodied struggle reflects Dewey's notion that excessive resistance disrupts aesthetic coherence; the singers' experiences of vocal problems were rarely fulfilling or harmoniously integrated into their artistic pursuits.

Vibrating matter and the transmateriality of singing

When reflux does not limit their vocal expression, the singers' experiences extend beyond internal physiological issues, appearing more varied and broader in scope than the previously presented reflux-challenged experiences. It can be said that the best singing experiences for the singers studied here range from the "cellular level" to the "universe." Here's an example of the former:

> I am bodily very present when I sing. The rhythm, melodies, different instruments, and message of the song seem to go to the cellular level and play me by starting already from there.
>
> (ID85–I)

Descriptions of the singers' best singing experiences included themes related to the body, body awareness, identity, mind and mental health, emotions, spirituality, power and attunement, limitlessness, space and time, and connection. The best singing experiences were associated with a strong sense of stimulation, such as energy, vitality, strength, empowerment, vigor, liveliness, flow, euphoria, exhilaration, and inspiration.

While singers with vocal problems found the loss of agency distressing, in the best singing experiences, agency was voluntarily relinquished, and the performance seemed to unfold effortlessly. Singers described how they didn't get tired, didn't have to work, and their bodies did not tense up. Overall, the activity felt easy and was accompanied by qualities such as flowing, floating, going smoothly, and traveling. The singer, the act of singing, sound, and music, emerged as intertwined actors, and the broader transactional configurations included elements such as sound waves, time, space, performance, and the self. The following response illustrates how singing can feel effortless and akin to "floating." Notably, the singer expresses their own agency implicitly using passive phasing, as did many other respondents of the study:

> Singing makes me happy, and it makes life feel meaningful. It's lovely just to float in the sound waves.
>
> (ID67–I)

In Dewey's terms, the balance of doing and undergoing in this example swings in favor of the latter. The following description shows how the body's material resistance can dissolve, making room for vocal flow. In such moments, the body's matter does not resist the passage of sound, allowing vibrations to travel without perceived obstacles. The vocal cords, which previously felt thick, painful, and limited during reflux episodes, no longer pose restrictions.

> The voice just flows — there are no boundaries, no sensations of pain. The voice is clear, and it doesn't get tired. [...] When singing goes smoothly, it feels like you don't have to 'work.' The larynx is open, and the vocal cords are not perceived (cf. when refluxed, the swelling and thickness are clearly felt).
>
> (ID81–I)

Unlike reflux experiences, where different body materials seemed to have conflicting orientations, in optimal singing experiences, the body as a material entity appears to work in unison. The body and mind were also united, and the singers felt oneness with their bodies, as illustrated in the following example:

> In that moment, I feel I am really one with my body. I'm happy and living in the moment, my worries are forgotten.
>
> (ID114–I)

Many responses described the body as unified rather than fragmented into discrete parts, as it often was during reflux symptoms. What contributes to this embodied wholeness? Singers cited transactors such as the flow of music, positive emotions, bodily activity, and the act of singing itself. However, one actor stood out above all: the vibration of sound, described as resonance, vibration, sound waves, and sonorousness.[11] Some singers mentioned how sound vibrations massage and nurture their bodies from within, with one stating that this "gives the body a sense of connection" (ID1-I). Vibrations were sometimes described as having power or physical force.[12]

> Singing can be felt in the entire body. In a way, the body vibrates with the power of singing.
>
> (ID42–I)

How does the vibration of sound contribute to a somaesthetic experience of a unified body? As discussed earlier, somaesthetic experiences can be transmaterial in nature. When perceived obstacles disappear, the singer's focus can shift from specific parts of the body to sensing transmaterial movements and vibrations throughout the body. By moving through tissues, vibration can draw the entire body into the same field of awareness, uniting it in the singer's experience as a cohesive whole. The effects of sound resonating through the whole body were illustrated by one of the singers like this:

> The whole body is resonating. The singing starts from the heels, from the ground, and travels through me. It's hard to describe in words, but it's like floating in a weightless space without any difficulty to stay in the moment or in the sound that resonates through me.
>
> (ID58–I)

In line with these observations, Nina Sun Eidsheim (2015) has highlighted that the materialities of singing involve not only the body's tissues but also their movement, particularly vibration. Vibrational movements blend the materialities of the body and sound (Eidsheim 2015: 1115).

In their best singing experiences, singers not only experienced unity within their bodies but also felt a strong connection across bodily boundaries with their environment, other people, or the world.[13] Singers also described being present in time, music, sound, resonance, and themselves. They characterized these experiences as being connected, involved in, one with, immersed in, or in the middle of something. Some singers' responses highlighted how singing can absorb them entirely, resulting in experiences where the body seems to disappear.[14]

> At its best, singing feels good in the entire body, and it actually feels like my body doesn't even exist but rather is one with the music.
>
> (ID90–I)

Eidsheim argues that perceived music is not merely an inner sensation of something external, as the vibrations of sound are palpably transmitted through the body, and the perceiver becomes one with the music in this vibrational process (Eidsheim 2015: 327, 606, 3357, 3491). Other previous research on voice and singing has also shown that vibrations and sounds can blur the boundaries between the inside and outside of the perceptual body (Silverman 1988: 80; Neumark 2010: xx; Chin 2011: 43; Tarvainen 2018b: 128). Sound artist and researcher Heidi Fast has explored how the human voice vibrates in both internal and external spaces, serving as an opening between different dimensions (Fast 2008: 346). Tiainen, for their part, has reflected on the connections between sound, movement, and matter, emphasizing that sound is movement—varying speeds, amplitudes, and waveforms. Sound connects different materials, including bodies, spaces, and technologies, through resonance and passage (Tiainen 2012: 141). In singing, the body's tissues move the surrounding air while the vibrations are transmitted back into the body, as Eidsheim has pointed out. The substance transmitting sound (e.g., air or water) and our body's reactions to it shape the singing experience (Eidsheim 2015: 1115). One of the singers responding to the survey described the phenomenon as follows:

> My whole body resonates, and the room is filled with the resonance of my body. I feel the music flowing through me.
>
> (ID88–I)

With its transmaterial capacities, the vibration of voice can connect bodies and subjects to one another (Neumark 2010; Fast 2016; Leppänen 2018: xx, xix), potentially even establishing a shared agency among them. In describing their best singing experiences, some respondents of this study illustrated agency as distributed in its nature rather than tied to a single actor.[15] Collective agency was especially prominent when singing in a choir.

> When performing, if the singing has gone well, I've had really euphoric experiences. In the whole body. It feels good, for example, when you sing with a choir completely in tune. It feels like we are one breathing, invincible animal.
>
> (ID40–I)

Singers with reflux may still experience positive moments of singing where they feel successful and connected in their performance. In such instances, their bodily sensations center on vibration and holistic awareness of the body rather than individual parts. The materialities of the body are perceived as a movement rather than as a static object, creating an all-encompassing singing experience. At its peak, such experiences lead singers to feel their bodily boundaries dissolve, merging with their surroundings or other elements. Transmaterial vibrations enable them to permeate conventional boundaries, resulting in feelings of disembodiment or collective agency with other people and the environment.

Transactional awareness

How can we further conceptualize the euphoric, flowing, vibrating, unified, and disappearing body experienced during the somaesthetic moments discussed above? This kind of body escapes traditional categorizations and, therefore, differs from anatomical models that depict organs as having fixed places and boundaries. According to Jaime Del Val and Stefan Lorenz Sorgner, anatomy is a body map that can only be viewed externally. It reflects a Cartesian division in which humans are seen as separate subjects, distinct from other subjects and objects in the world. Del Val and Sorgner propose deconstructing this division through the development of perceptual technologies and sensory organs using the concept of the *postanatomical body* (del Val and Sorgner 2011: 2–3).

In somaesthetics, the anatomical or physiological body and the related human disconnection from the rest of the world are dissolved by cultivating somaesthetic experiences in practice.[16] The soma described in somaesthetics does not necessarily correspond to the physiological body as defined by medical science. The aim is not to outline a general model of the body but to explore a variety of perceived bodies. Somatic awareness can take widely different forms depending on the situation and cultural context (Shusterman 2011: 317–318).

The body becomes co-constituted with music differently than in "everyday experience."[17] It is shaped not as a physical object but as movement, flow, and vibration. When the heavy and resistant body familiar from everyday experience vanishes from awareness, the singer may feel their body disappearing. In such moments, the everyday body no longer co-constitutes within the transactions the body is engaged in. Earlier in this chapter, I referred to Dewey's idea of the materials of the world becoming the medium of artistic expression. The tissues of a singer's body and the molecules of air, which were simply everyday matter moments before, can instantly transform during a performance. The walls of the space cease to be mere silent, passive barriers and become resonating, sound-shaping elements that vibrate alongside the singer. Similarly, the singer's muscles are no longer just tissues maintaining bodily form; they become the matter that shapes the character of vocal vibration.

Dewey notes that sometimes a material component of a work of art, such as paint in a painting, may not integrate seamlessly into the whole. He suggests that even great artists do not consistently achieve this *sense of connection* in their work (Dewey 2005/1934: 299). This can also happen with vocal problems, when the expressive potential of the body is lost, leaving it fragmented—even if the intent was to transform it into a cohesive medium of musical expression. In this sense, one respondent's description of music or singing as an "escape from everyday life" is fitting, even from a material standpoint:

> At best, it [singing] is a kind of utopian space, an escape from everyday life but connected to everyday life.
>
> (ID125–I)

In peak singing experiences, the singer may lose the sense of self as a distinct entity and feel themselves as a part of a greater whole, where the subject/object divide becomes blurred. I suggest that these perceptions represent *transactional awareness*, wherein the singer gains an ontological insight into their oneness with the world. In such moments, the singer becomes aware of their entanglement with their environment, participating in living transactions that co-constitute them as singing soma. The singer no longer perceives themselves as "a-singer-as-a-whole" but as "a-singer-in-the-environment-as-a-whole" (cf. Dewey and Bentley 1949: 109)—or even "a-singer-in-the-universe-as-a-whole," as the following survey response illustrates:

> It's as if time and boundaries would disappear and none of the worries and pain of the world could touch you. Poetically, I could say that it's like being one with the universe.
>
> (ID53–I)

According to Leddy, Dewey's thinking suggests that works of art help us realize our belonging to the universe. An aesthetically intense object can even evoke religious experiences. Through art, we can sense our world beyond everyday life and be "carried out beyond ourselves," as Dewey states. Such perceptions enable us to move past our usual boundaries and feel a sense of belonging to a broader whole. While some interpret these ideas as a departure from Dewey's naturalistic approach, Leddy sees it still rooted in materialism, viewing humans and the world as equally material phenomena (Dewey 2005/1934: 202–203; Leddy 2019: 46).

Laasonen Belgrano describes peak singing experiences as moments where "spirituality" emerges as a contradictory, disruptive, frightening, and seductive force. She captures how vocal sound production is a process of meaning-making and a quest for ever-deeper knowledge of life and living (Laasonen Belgrano 2018). Such descriptions might seem poetic or mystical to non-singers, but for singers, these experiences are often profoundly embodied, not merely aspirational. The aliveness felt in singing can surpass most everyday experiences. Laasonen Belgrano's insights reflect the intra- or transactional nature of singing, wherein co-constitutive processes reach far and wide, encompassing "knowing about life and living" (Laasonen Belgrano 2018).

Even if a singer is ontologically one with their environment, there are moments when they perceive themselves as distinct from their surroundings. For example, when practicing, the singer may adopt an instrumental view of their body, treating it as an object. Similarly, during vocal problems, the unity of body and mind often breaks, and the body becomes objectified. Dewey and Bentley note that there are contexts where it is beneficial to view the organism as separate from its environment or even to see its different parts as separate from each other. This approach can be necessary when identifying specific material factors contributing to a disease or symptom. For example, they illustrate this point with issues involving the digestive system, where its function may be impaired and require focused intervention (Dewey and Bentley 1949: 323). The instrumentalization of the body in medical contexts and singing practice often serves this kind of goal, informed by a physiological perspective.

Conclusions

In this chapter, I have explored singers' perceptions of materialities in reflux-induced vocal problems and somaesthetic singing experiences. *Experiences of vocal problems* are often shaped by uncomfortable and even painful embodied sensations. Excess materials and altered tissues induced by reflux can occupy space, hindering the resonation of the voice in the body's cavities and disrupting the singer's agency. This fragmented sense of agency often brings about complex emotions such as fear, sadness, frustration, and anger. Excessive resistance can fragment the experience into discrete parts, isolating sensations in specific tissues and causing the singer to lose their sense of being at the center of their experience.

In contrast, *somaesthetic experiences* evoke sensations of vibration, wholeness, and even disembodiment. Instead of struggling to maintain agency, the singer may willingly surrender to the "power of singing." In such moments, the singer's attention shifts from perceiving the body as a heavy, static object to engaging with the permeating vibrations, movements, and fluidities of bodily (and other) matter. Transmaterial vibrations dissolve the boundaries of matter by moving different materials in unison and uniting them in the experience. The singer's focus is no longer on specific body parts or tissues. Instead, the entire body seems to "flow" or "vibrate" as one. Even the surrounding space can join in this collective agency of materials, leading the singer to lose their sense of bodily boundaries and merge with the world. In this way, the singer approaches a state of transactional awareness, realizing their intertwinement and co-constitution with the world.

The singer's body, as a vocal soma, is co-constituted in the act of singing either as object-like matter or as a movement—though rarely as both simultaneously. When examining what co-constitutes in "successful" or "failed" singing, we observe different transactional events producing different bodies and experiences. It is not merely the vibration of sound and movement of matter that change but also the way they are perceived. Therefore, "successful" and "failed" singing are also culturally defined experiences, influenced by the aesthetic ideals adopted by the singer.

As previously noted, some participants in this study described "losing" their bodies during their best singing experiences, with the "body of everyday experience" seemingly disappearing. This raises the question of whether we have become so accustomed to perceiving our bodies as heavy, resistant, and uncomfortable in daily life that when these sensations vanish, our familiar bodily perception also "disappears" from our experience. If embodied states of effortlessness, flow, and wholeness are largely absent from our day-to-day experience, it may be time to reevaluate the culturally organized activities, habits, environments, and technologies that shape how we perceive ourselves. Integrating somaesthetic experiences into daily life is essential, and the human voice, with its palpable corporeality and sociality, serves as a valuable tool for this purpose. Indeed, vocality has long been embedded in many cultures' daily activities, as evidenced by communal vocal expressions during work and celebration. As Virginie Magnat (2019: xiv) suggests, vocality should be recognized as "a vital source of embodied knowledge, creativity, and well-being, grounded in process, practice, and place, as well as a form of social and political agency." Somaesthetically attuned vocal practices could build on Magnat's insights, incorporating, for example, Eidsheim's vibratory approach, Fast's vocal practice for developing sensibility, or my own Voicefulness® method, which focuses on fostering body awareness through voice (Eidsheim 2015: 698; Fast 2016; Tarvainen 2022). These approaches share a common objective: to move beyond performance-driven aesthetic pursuits by centering embodied vocal experience, thereby making somaesthetic experiences more widely accessible.

When asked about their worst and best singing experiences, the singers in this study did not solely discuss their voices or vocal ideals. They described their past, future, dreams, identity, feelings, thoughts, embodied experiences, and connections with themselves, others, and the world. This wide spectrum of reflections underscores the rich, complex, and far-reaching nature of vocal transactions. The data revealed singing as an experience that encompasses the full range of being human, even touching on the ontology of existence. In this light, vocal problems take on a more profound meaning, significantly impacting a singer's life and well-being.

Notes

1 "The Feel of Singing: Aesthetic Body Awareness in the Vocal Experiences of Deaf, Tone Deaf, and Vocally Disordered Singers" was a three-year postdoctoral research project (2018–2021) funded by the Kone Foundation. It consisted of three sub-studies, of which I will discuss the data related to singers with vocal disorders. The text at hand and included analysis have been further developed and modified from a previous Finnish article published in the *Musiikki* journal, a periodical of The Finnish Musicological Society (Tarvainen 2021).

2 Reflux occurs in two types: gastroesophageal reflux disease (GERD), commonly known as heartburn, and laryngopharyngeal reflux disease (LPR), which can cause symptoms such as sore throat, globus sensation, hoarseness, throat clearing, chronic cough, and asthma (Koufman et al. 1996: 215; Lipan et al. 2006: 261–263).

3 The similarities and differences between new materialism and somaesthetics are explored in an essay by Wojciech Małecki and Simon Schleusener (2015), which discusses Gilles Deleuze's philosophy, a key influence on many new materialisms, and Richard Shusterman's somaesthetic philosophy.

4 Dewey and Bentley's *transaction* parallels Barad's (2007) *intra-action* developed in post-humanist and new materialist frameworks. The resemblances have previously been discussed by Fischer (2018), Hammarström (2010), Reynolds (2019:

131–132), and Rosiek (2013: 700; 2018), among others. Fischer (2018) has criticized the fact that the concept of intra-action has been developed without taking into account Dewey and Bentley's corresponding concept.

5 The term *co-constitution* has also been used in somaesthetics (Kestenbaum 1977; Aldrich 2008, according to Antony 2015; Rosiek 2013; Sullivan 2001). A parallel concept is also used in new materialist thinking (Barad 2007; Juelskjaer and Schwennesen 2012).

6 For more on Dewey's theory of agency, see, e.g., Östman and Öhman 2010.

7 Inspired by Rimmer's (2020) ideas on transactional analysis, I previously developed and tested a method to examine the transactions between human voices and background music in service industry workplaces as part of the ACMESOCS research project funded by the Research Council of Finland (Tarvainen 2023b). In the current study, I applied a streamlined version of this method.

8 A direct quote from my research data, with the respondent's identification number.

9 This singer had also been diagnosed with dystonia, which she reports worsens with reflux.

10 Even the entire act of "singing during reflux" can appear as an actor within complex transactional configurations, similar to findings in my prior analyses on the transactions of human voice and background music (Tarvainen 2023b).

11 Vibration was also a significant experiential quality for these singers, though other qualities were also present, such as freedom, lightness, ease, naturalness, fluency, movement, speed, flexibility, firmness, warmth, growth, rootedness, supportiveness, cleansing, healing, sensitivity, sensuality, orgasticity, and openness.

12 In Dewey's soft naturalism, which does not include an immaterial dimension beyond matter (Määttänen, 2015a: 86), transmateriality, motion, vibration, vitality, energy, and power do not require any "higher" forces to exist.

13 Dewey and Bentley's transactional approach moves beyond the idea of human skin-boundedness, viewing perception and behavior as *transdermal* phenomena (Phillips 2013: 97).

14 In addition to the disappearing body, singers often described the absence of elements from their everyday experience, such as boundaries, self, time, pain, or worries.

15 Bennett (2010) discusses distributive agency in the context of new materialist frameworks.

16 The need for sensory training has also been articulated in new materialist thought (Connolly 2010: 186–187, 193).

17 These differently co-constituted experiences can emerge within everyday actions and practices as well. They do not necessarily require special occasions or an artistic framework to be realized.

References

Aldrich, Rebecca M. 2008. "From Complexity Theory to Transactionalism: Moving Occupational Science Forward in Theorizing the Complexities of Behavior." *Journal of Occupational Science* 15 (3): 147–156. https://doi.org/10.1080/14427591.2008.9686624

Alhanen, Kai. 2013. *John Deweyn kokemusfilosofia*. Helsinki: Gaudeamus.

Antony, Alexander. 2015. "Tacit Knowledge and Analytic Autoethnography: Methodological Reflections on the Sociological Translation of Self-Experience." In *Revealing Tacit Knowledge. Embodiment and Explication*, edited by Frank Adloff, Katharina Gerund, and David Kaldewey, 139–168. Bielefeld: transcript Verlag.

Barad, Karen. 2007. *Meeting the Universe Halfway: Quantum Physics and the Entanglement of Matter and Meaning*. Durham and London: Duke University Press.

Bennett, Jane. 2010. *Vibrant Matter: A Political Ecology of Things*. Durham & London: Duke University Press.
Bolt, Barbara. 2013. "Introduction: Toward the New Materialism through the Art." In *Carnal Knowledge: Towards a "New Materialism" through the Arts*, edited by Estelle Barrett and Barbara Bolt. London & New York: I. B. Tauris.
Brinkmann, Svend. 2011. "Dewey's Neglected Psychology: Rediscovering His Transactional Approach." *Theory & Psychology* 21 (3): 298–317. https://doi.org/10.1177/0959354310376123
Chin, Christina. 2011. "Vibrate ... Resonate ... Quicken the Educational Experience into Intensest Life." *Art Education* 64 (3): 41–45. https://doi.org/10.1080/00043125.2011.11519127
Connolly, William E. 2010. "Materialities of Experience." In *New Materialisms: Ontology, Agency, and Politics*, edited by Diana Coole and Samantha Frost, Kindle edition, 178–200. Durham & London: Duke University Press.
del Val, Jaime, and Stefan Lorenz Sorgner. 2011. "A Metahumanist Manifesto." *The Agonist (Nietzsche Circle)* IV (2): 1–4.
Dewey, John. 2005/1934. *Art as Experience*. Kindle edition. New York: Penguin Publishing Group.
Dewey, John, and Arthur F. Bentley. 1949. *Knowing and the Known*. Boston: The Beacon Press.
Dolphijn, Rick, and Iris van der Tuin. 2012. "'Matter Feels, Converses, Suffers, Desires, Yearns and Remembers': Interview with Karen Barad." In *New Materialism: Interviews & Cartographies*, edited by Rick Dolphijn and Iris van der Tuin, 48–70. Ann Arbor, MI: New Humanities Press.
Eidsheim, Nina Sun. 2015. *Sensing Sound: Singing & Listening as Vibrational Practice*. Kindle edition. Durham and London: Duke University Press.
Fast, Heidi. 2008. "An Emergent Tuning as a Molecular Organizational Mode." *Inflexions* (4): 342–357. https://www.inflexions.org/n4_An-Emergent-Tuning-as-an-Organizational-Molecular-Mode-by-Heidi-Fast.pdf
Fast, Heidi 2016. "Singing Touches Shared Sensibility." *Boldness Blog, Kone Foundation*. https://koneensaatio.fi/en/stories/singing-touches-shared-sensibility/
Ferrando, Francesca. 2012. "Towards a Posthumanist Methodology. A Statement." *Frame: Journal for Literary Studies* 25 (1): 9–18. https://doi.org/10.1080/13825577.2014.917005
Fischer, Clara. 2018. "Revisiting Feminist Matters in the Post-Linguistic Turn: John Dewey, New Materialisms, and Contemporary Feminist Thought." In *New Feminist Perspectives on Embodiment*, edited by Clara Fischer and Luna Dolezal, 83–102. https://doi.org/10.1007/978-3-319-72353-2_5
Hammarström, Matz. 2010. "On the Concepts of Transaction and Intra-Action." In *The Third Nordic Pragmatism Conference*, 1–15. Uppsala.
Jones, Tamsin. 2016. "Introduction: New Materialism and the Study of Religion." In *Religious Experience and New Materialism: Movement Matters*, 1–23. Hampshire, UK & New York, NY: Palgrave Macmillan.
Juelskjaer, Malou, and Nete Schwennesen. 2012. "Intra-Active Entanglements: An Interview with Karen Barad." *KVINDER, KØN & FORSKNING* 21 (1–2): 10–23.
Kestenbaum, Victor. 1977. *The Phenomenological Sense of John Dewey: Habit and Meaning*. Atlantic Highlands, New Jersey: Humanities Press.
Koufman, James, Robert T. Sataloff, and Robert Toohill. 1996. "Laryngopharyngeal Reflux: Consensus Conference Report." *Journal of Voice* 10 (3): 215–216. https://doi.org/10.1016/s0892-1997(96)80001-4
Koufman, James A., Milan R. Amin, and Marguerite Panetti. 2000. "Prevalence of Reflux in 113 Consecutive Patients with Laryngeal and Voice Disorders." *Otolaryngology – Head and Neck Surgery* 123 (4): 385–388. https://doi.org/10.1067/mhn.2000.109935

Laasonen Belgrano, Elisabeth. 2018. "Ornamenting Vocality: An Intra-Active Methodology for Vocal Meaning-Making." *Ruukku – Studies in Artistic Research* (9). https://www.researchcatalogue.net/view/370801/370802

Laasonen Belgrano, Elisabeth 2021. "Mapping the Burden of Vocality: French 17th Century Vocal Lamentations, Japanese Meditation and Somatic Intra-Action." In *Somatic Voices in Performance Research and Beyond*, edited by Christina Kapadocha. Routledge Voice Studies: Research Monographs. Abingdon, Oxon & New York, NY: Routledge.

Lechien, J. R., A. Schindler, C. Robotti, L. Lejeune, and C. Finck. 2019. "Laryngopharyngeal Reflux Disease in Singers: Pathophysiology, Clinical Findings and Perspectives of a New Patient-Reported Outcome Instrument." *European Annals of Otorhinolaryngology, Head and Neck Diseases* 136 (3): S39–S43. https://doi.org/10.1016/j.anorl.2018.08.008

Leddy, Thomas. 2019. "Dewey, Materiality, and the Role of the Visual (Studio) Arts in the Liberal Arts." *The Journal of Aesthetic Education* 53 (4): 40. https://doi.org/10.5406/jaesteduc.53.4.0040

Lennon, Kathleen. 2019. "Feminist Perspectives on the Body." In *Stanford Encyclopedia of Philosophy*, edited by Edward N. Zalta, Fall 2019. Stanford: Metaphysics Research Lab, Stanford University. https://plato.stanford.edu/archives/fall2019/entries/feminist-body/

Lenti, Marco Vincenzo, Giovanni Cammarota, Francesca Vidali, Giovanna Masala, Benedetta Bendinelli, Giovanni Gasbarrini, Gino Roberto Corazza, and Antonio Di Sabatino. 2019. "Reflux Symptoms in Professional Opera Soloists." *Digestive and Liver Disease* 51 (6): 798–803. https://doi.org/10.1016/j.dld.2018.11.026

Leppänen, Taru. 2018. "Always More than Two: Vibrations, the Foetus, and the Pregnant Person in Childbirth Singing Practices." *NORA – Nordic Journal of Feminist and Gender Research* 26 (2): 99–111. https://doi.org/10.1080/08038740.2018.1462255

Leppänen, Taru, and Milla Tiainen. 2018. "Trans-Becomings in Western 'Classical' Singing: An Intra-Active Approach." *Ruukku – Studies in Artistic Research* (9). https://www.researchcatalogue.net/view/372940/372941

Lipan, Michael J., Joy S. Reidenberg, and Jeffrey T. Laitman. 2006. "Anatomy of Reflux: A Growing Health Problem Affecting Structures of the Head and Neck." *Anatomical Record – Part B New Anatomist* 289 (6): 261–270. https://doi.org/10.1002/ar.b.20120

Lopes, Leonardo Wanderley, Gyllyane Furtado Cabral, and Anna Alice Figueiredo De Almeida. 2015. "Vocal Tract Discomfort Symptoms in Patients with Different Voice Disorders." *Journal of Voice* 29 (3): 317–323. https://doi.org/10.1016/j.jvoice.2014.07.013

Macpherson, Ben. 2021. "The Somaesthetic in-between: Six Statements on Vocality, Listening and Embodiment." In *Somatic Voices in Performance Research and Beyond*, edited by Christina Kapadocha. London: Routledge.

Magnat, Virginie. 2019. *The Performative Power of Vocality*. New York: Routledge.

Małecki, Wojciech, and Simon Schleusener. 2015. "'What Affects Are You Capable of?' On Deleuze and Somaesthetics." In *Deleuze and Pragmatism*, edited by Sean Bowden, Simone Bignall, and Paul Patton, 216–234. New York & London: Routledge.

Merati, Albert L. 2020. "Reflux and the Performer's Voice." In *The Performer's Voice*, edited by Michael S. Benninger, Thomas Murry, and Michael M. Johns, 2nd ed., 193–208. San Diego, CA: Plural Publishing, Inc.

Määttänen, Pentti. 2015a. "Emotionally Charged Aesthetic Experience." In *Aesthetics and the Embodied Mind: Beyond Art Theory and the Cartesian Mind-Body*

Dichotomy, Contributions to Phenomenology 73, edited by Alfonsina Scarinzi, 85–99. Dordrecht: Springer. https://doi.org/10.1007/978-94-017-9379-7_6

Määttänen, Pentti. 2015b. *Mind in Action: Experience and Embodied Cognition in Pragmatism. Studies in Applied Philosophy, Epistemology and Rational Ethics, Volume 18*. Cham, Heidelberg, New York, Dordrecht, London: Springer. https://doi.org/10.1007/978-3-319-17623-9

Määttänen, Pentti. 2017. "Emotions, Values, and Aesthetic Perception." *New Ideas in Psychology* 47: 91–96. https://doi.org/10.1016/j.newideapsych.2017.03.009

Nacci, Andrea, Luca Bastiani, Maria Rosaria Barillari, Massimo Martinelli, Jerome R Lechien, Federica Simoni, Stefano Berrettini, and Bruno Fattori. 2020. "Reflux Symptom Index (RSI) and Singing Voice Handicap Index (SVHI) in Singing Students: A Pilot Study." *Journal of Voice*. 1–10. https://doi.org/10.1016/j.jvoice.2020.05.005

Neumark, Norie. 2010. "Introduction: The Paradox of Voice." In *VOICE: Vocal Aesthetics in Digital Arts and Media*, edited by Norie Neumark, Ross Gibson, and Theo van Leeuwen, xv–xxxiii. Cambridge & London: The MIT Press.

Östman, Leif, and Johan Öhman. 2010. "A Transactional Approach to Learning." Presented at the John Dewey Society, AERA Annual Meeting, Denver, Colorado.

Phillips, Trevor J. 2013. *Transactionalism: An Historical and Interpretive Study*, edited by John Patterson and Kirkland Tibbels. Ojai, CA: Influence Ecology.

Reynolds, Daniel. 2019. *Media in Mind*. Oxford: Oxford University Press.

Rimmer, Mark. 2020. "Trans-Actions in Music." In *John Dewey and the Notion of Trans-Action a Sociological Reply on Rethinking Relations and Social Processes*, edited by Christian Morgner, Kindle edition, 2770–3550. Hampshire, UK & New York, NY: Palgrave Macmillan.

Rosiek, Jerry Lee. 2013. "Pragmatism and Post-Qualitative Futures." *International Journal of Qualitative Studies in Education* 26 (6): 692–705. https://doi.org/10.1080/09518398.2013.788758

Rosiek, Jerry Lee 2018. "Art, Agency, and Inquiry: Making Connections between New Materialism and Contemporary Pragmatism in Arts-Based Research." In *Arts-Based Research in Education: Foundations for Practice*, 2nd ed., edited by Melisa Cahnmann-Taylor and Richard Siegesmund, 2, 32–47. New York: Routledge.

Ryder, John. 2014. "Experience and Judgement: Political and Aesthetic." In *Beauty, Responsibility, and Power: Ethical and Political Consequences of Pragmatist Aesthetics*, edited by Leszek Koczanowicz and Katarzyna Liszka, 63–76. Amsterdam and New York: Rodopi.

Shusterman, Richard. 1994. "Dewey on Experience: Foundation or Reconstruction?" *Philosophical Forum* XXVI (2): 127–148.

Shusterman, Richard. 2000. *Pragmatist Aesthetics: Living Beauty, Rethinking Art*. 2nd ed. New York: Rowan & Littlefield.

Shusterman, Richard. 2008. *Body Consciousness: A Philosophy of Mindfulness and Somaesthetics*. Cambridge: Cambridge University Press.

Shusterman, Richard. 2011. "Soma, Self, and Society: Somaesthetics as Pragmatist Meliorism." *Metaphilosophy* 42 (3): 314–327. http://onlinelibrary.wiley.com/doi/10.1111/j.1467-9973.2011.01687.x/full

Shusterman, Richard. 2012a. "Body and the Arts: The Need for Somaesthetics." *Diogenes* 59 (1–2): 7–20. https://doi.org/10.1177/0392192112469159

Shusterman, Richard. 2012b. *Thinking through the Body: Essays in Somaesthetics*. Cambridge: Cambridge University Press.

Silverman, Kaja. 1988. *The Acoustic Mirror: The Female Voice in Psychoanalysis and Cinema*. Bloomington: Indiana University Press.

Sullivan, Shanon. 2001. *Living across and through Skins: Transactional Bodies, Pragmatism, and Feminism*. Bloomington & Indianapolis: Indiana University Press.

Suzuki-Bosco, Noriko. 2016. "Gaston Bachelard's 'Material Imagination' + Karen Barad's New Materialism." Postgraduate Research, Winchester School of Art. https://blog.soton.ac.uk/wsapgr/2016/04/04/gaston-bachelards-material-imagination-karen-barads-new-materialism/.

Tarvainen, Anne. 2016. "Vokaalinen soomaestetiikka: Kehotietoisuuden esteettiset mahdollisuudet ihmisen äänenkäytössä ja kuuntelemisessa." ["Vocal Somaesthetics. The Potentials of Body Awareness in Voice and Listening." *The Finnish Yearbook of Ethnomusicology* 28: 1–39. http://etnomusikologia.journal.fi/article/view/60239/21141

Tarvainen, Anne. 2018a. "Democratizing Singing: Somaesthetic Reflections on Vocality, Deaf Voices, and Listening." *Pragmatism Today* 9 (1): 91–108. http://www.pragmatismtoday.eu/summer2018/Democratizing-Singing-Somaesthetic-Reflections-on-Vocality-Deaf-Voices-and-Listening-Anne-Tarvainen.pdf

Tarvainen, Anne 2018b. "Singing, Listening, Proprioceiving: Some Reflections on Vocal Somaesthetics." In *Aesthetic Experience and Somaesthetics*, edited by Richard Shusterman, 120–142. Leiden & Boston: Brill. https://urn.fi/URN:NBN:fi:tuni-202105104704

Tarvainen, Anne. 2021. "Vastustava, värähtelevä keho: kehon materiaalisuuksia refluksia sairastavien laulajien soomaesteettisissä kokemuksissa." ["Resisting, Vibrating Body: Materialities of the Body in the Somaesthetic Experiences of Singers with Acid Reflux." *Musiikki* 51 (1): 9–43. https://musiikki.journal.fi/article/view/107355

Tarvainen, Anne 2022. "Oma ääni rauhoittaa, hoitaa ja luo yhteyden kokemuksia Voicefulness- kursseilla." ["Own voice soothes, nurtures and creates experiences of connection in Voicefulness courses."] *Voicefulness®-blog*. https://www.voicefulness.fi/post/voicefulness-kokemuksia-kursseilta

Tarvainen, Anne. 2023a. "How to Apply Somaesthetics?: Practices, Methods, and Research Design in Somaesthetic Approach." *The Journal of Somaesthetics* 9 (1/2): 108–129. https://doi.org/10.54337/ojs.jos.v9i1/2.7891

Tarvainen, Anne. 2023b. "Musiikki hukkuu puheensorinaan: Taustamusiikki ja ihmisääni toimijoina palvelualojen työpaikkojen ääniympäristöissä." ["Music Drowns in Chatter: Background Music and Human Voice as Actors in the Sound Environments of Service Sector Workplaces."] *The Finnish Yearbook of Ethnomusicology* 35: 207–233. https://doi.org/10.23985/evk.127849

Thomaidis, Konstantinos, and Ben Macpherson. 2015. *Voice Studies: Critical Approaches to Process, Performance and Experience*, edited by Konstantinos Thomaidis and Ben Macpherson. London & New York: Routledge.

Tiainen, Milla. 2012. *Becoming-Singer: Cartographies of Singing, Music-Making and Opera*. PhD dissertation, University of Turku.

Tiainen, Milla. 2013. "Revisiting the Voice in Media and as Medium: New Materialist Propositions." *NECSUS. European Journal of Media Studies* 2 (2): 383–406. https://doi.org/10.5117/NECSUS2013.2.TIAI

Tiainen, Milla. 2017. "Sonic Technoecology: Voice and Non-Anthropocentric Survival in the Algae Opera." *Australian Feminist Studies* 32 (94): 359–376. https://doi.org/10.1080/08164649.2017.1466651

5 Caring, touching, voicing

On how voice comes to voice in academia

Monika Rogowska-Stangret and Malou Juelskjær

Introduction: Why we started thinking about and with a voice in academia

This chapter is part of ongoing conversations between the authors since 2015 about living a feminist academic life in a neoliberal academia, that is, a form Western academia has taken in late capitalism.[1] We have shared our experiences of teaching, researching, writing, collaborating, organizing, applying for funding, and living, while located in different academic contexts: in Northern and Eastern Europe. Acceleration in academia and the heated discussion about slow and fast science inspired our previous paper on how different temporalities and velocities impact our lives as scholars (Juelskjær and Rogowska-Stangret 2017). Intense periods of teaching directed our attention to how we use our voices – not only when teaching but during other academic activities: when presenting at conferences, taking the floor during research seminars, giving feedback to students and colleagues, leading discussions, chatting or networking, but also when listening to others talk, lecture, or otherwise verbally interact. We decided to dwell for a moment on the question of voice in academia and this is how it came into focus in our joint work. The questions proliferated: How does academia sound from our situated perspectives? What is going on with voice in academia? What voices does academia re/produce, ex/in/clude, ab/use? What different "layers" of voices are affected (physiological, organic, cultural, social, individual, sensorial)? What are considered professional academic voices? Which voices are considered too quiet, too loud, too low, too high, too dry, too uncontrolled, too slow, too fast, stuttering, lisping, muttering – and in which situations?[2]

Indeed, the voice does not seem an accidental or arbitrary topic when researching academic life. Spoken language is among the milieus we, as academics, most commonly inhabit. Even more than that: the voice is *of* academia, meaning that speaking-voicing is intricately entangled with (for example) the flows and blockages mobilized by power relations operating in the academic environment. Being *of* academia, voice is formed by experiences of being an academic, doing academic work and caring about academia – and, voice reshapes those experiences. In the act of speaking-voicing, a worlding of an

DOI: 10.4324/9781003390992-8

academic life takes place: the ways in which voice comes to voice in academia are instances of world-making. As such, voice and how to voice both concerns us and piques our curiosity. As scholars, we are invested in the "how" of practicing *response-ability* (Haraway 2008) within academia. Following Haraway and joining others (activists and scholars), we consider *response-ability* as an ethics of *how* to practice responsibility through enabling and being open to response, and to find ways to tune into the multitude of ways that the "surround" (involving humans, more-than-, or other-than-humans) may seek to respond. An ethics of *response-ability* further requires the researchers to tune in to – or to be in touch with – those responses to enable the research to contribute to more just world-makings. In what follows, we thus also want to investigate the ethics of voice in academia: how to touch response-ably with one's voice? How to make dis/connections with one's voice ethically? What is the quality of voice-touch (touching voice) we wish to address?

Within the context of this chapter, we zoom in and attend to the onto-politics of living an academic life through micro-moments which comprised acts of voicing. Thinking with our own too quiet, too loud, too low, too high, too dry, too uncontrolled, too slow, too fast, stuttering, lisping, muttering voices, we turn to questions of ethics as care and caring. It is a matter of caring for one's voice while also, at the same time, caring for those that hear it. This relational aspect of voice reminds us of the important work on the intersubjectivity and reciprocity of voice as a phenomenon that opens a relation between the one who speaks and the one who listens, which is offered by Adriana Cavarero (2005). One way of caring for one's voice and for those listening to it is to consult a voice coach. Consequently, in this chapter, we reflect on three experiences with three different voice coaches (a male and two female coaches) whom we consulted individually in our home cities as part of an academic support system or academic training and on a conversation with Joanna Halszka Sokołowska, a musician and voice coach.[3] Our reflections are based on personal notes, teaching materials, and audio recordings. Our aim is not to perform a detailed analysis of our experiences; instead, we dwell on selected moments from the coaching sessions. Voice coaching sessions are very particular practices. However, following definitions of voice after scholars who approach voice through feminist new materialist thinking, voice coaching sessions as part of academic training and support are relevant to understanding what goes on with a voice in academia: "as a bodily and sonic phenomenon, the voice is already, in itself, a result of specific corporeal, social and mental techniques" (Tiainen 2017: 364). The particular coaching moments enabled the opening of a space for sensing what is going on with voice in academia. Fathomed from our situated perspectives, they point to concrete voice phenomena (e.g., the question of "professional voice," the pitch of a voice, and "mistakes" made while speaking-vocalizing) that enable us to reflect on specific academic practices, like teaching, and the ethics involved in them. The three voice coaching sessions (and a conversation) also helped to structure this chapter, providing three main foci for our reflections:

1 voice and trans-corporeality: how voice is not a neutral mediator
2 response-ability and touch: working on your pitch during voice coaching
3 voicing as touching the not-yet

In each section, we unfold the selected moments through a narrative of a first-person (also visually by using italics), an "I," who is engaging in situations of voice coaching, and who is simultaneously engaged in "inner voice considerations" of the situation.

Rather than offering solutions to problems, the specific coaching situations allowed us to "stay with the trouble" (Haraway 2016), materialized with and in how our voices were coming to voice while pondering on the possible entangled effects of those occasions. By staying with the trouble, Haraway argues to neither ignore nor give in/up but rather to search for ways of "making-with" (sympoiesis) the already complex entanglements, to find ways of building more liveable futures. Furthermore, we use voice coaching encounters to reconsider ethics and care in relation to voice in academia in an attempt at "embracing the tension between the concrete and the speculative" (Puig de la Bellacasa 2017: 95). In this view, speculation involves to not let "a situation or a position – nor even the acute awareness of pervasive dominations – define in advance what is or could be" (Puig de la Bellacasa 2011: 96). This tension is productive; it may create a space of tinkering with questions of care. It brings to the fore concrete, situated experiences of discomfort. In our take, the concrete voice coaching practices of care allow us to speculate on (without, however, determining) the ways in which a liveable and caring academic world might exist. We want to contribute to debates on voices in academia, seeing our chapter as a modest contribution to a rich and flourishing field of voice studies, of which a – broadly understood – feminist new materialism is but one among many approaches (Tiainen 2013; Mazzei and Jackson 2017; Neumark 2017; Moisala et al. 2017; Fast and Tiainen 2018; Leppänen 2018; Tiainen 2017).

Before analyzing the three focuses listed above, we briefly situate our thoughts on voice within academia. The question of voice in academia, especially from a feminist perspective, is already a broad topic of research that extends beyond our concerns in this chapter. However, we would like to point to chosen approaches to investigating voice in academia as an important part of feminist scholarship. This will further help us to introduce guiding questions that shape how we think about and with a voice in academia.

Voice and/in academia as a feminist issue

Voice is a multifaceted phenomenon and, in feminist archives, the question of voice in academia is oftentimes investigated as linked to questions of power relations and accessibility, such as: whose voice is legitimized as a voice of an academic? Who has the right to speak in the space of the university? And whose voices are deemed inadequate? When Virginia Woolf encouraged women to think, she also pointed to a construct of "a monstrous man," who is strong, "loud of voice," and obsessive about demarcating the limits of a human (1938: 160). In the context of academia, being "loud of voice" is a clear sign that the academic space is where the person (he) feels at home. This contrasts with women-identified scholars, who sometimes ponder: was it "something in my voice, my attitude, my tension [that] 'marked' that I did not feel completely 'in my place,' and authorized the reaction that signalled to me that I wasn't, indeed" (Stengers and Despret 2014: 96)? A sounding voice – a particularly vulnerable instrument – reveals here deeper layers of conflicting power relations that participate in constituting an individual woman-identified academic. Mary Beard throws light on this matter by pointing out that using one's voice in public was an attribute defining manhood from Ancient times. Thus, a female voice appearing publicly is always disorienting and is not identified with an authority figure (2017).

The insecurity about one's voice and feeling not "in my place" are organized through, for example, "politics of citation," understood here, after Ahmed, as "a [...] reproductive technology, a way of reproducing the world around certain bodies" (2013). This reproductive technology wipes certain bodies out: erases, deforms, unifies, silences, or marginalizes their voices (Todd 2016; Rogowska-Stangret and Cielemęcka 2020). This wiping out does not concern only voices in writing but also in the very acts of listening to vocal utterances and in creating spaces/opportunities that enable speech at all. Indeed, "relationality can be effected through sound" (Tiainen 2017: 361).

That is why, when thinking about voice in academia as a feminist issue, we are particularly drawn to the words of Claire Colebrook: "Feminism is always the question of who: who speaks, for whom and whose subjectivity is presupposed in the grammar of the question?" (2017: 9). This important question can support our reflections concerning the event of speaking in the academy. "Who speaks" (Why they? Who bestows the right to speak? Who is allowed a voice? Who does not speak while we speak?), "for whom" (How do we construct the figure of a listener? What does it mean to listen/hear? What do we want to achieve by addressing this particular audience in this particular way?). Finally, we wish to explore "whose subjectivity is presupposed in the grammar of the question" (What is the structure of the language used? How voluminous/polyphonic is the language in terms of voices? Is it "my" voice? Are these "my" words? How many voices do I allow to resonate in "my own" speech? What power relations do I recreate while speaking – consciously and unconsciously, on purpose and by accident?).

All these questions are vibrantly present whenever someone opens their mouth in acts of speaking and listening in academia. Or, that is what this chapter claims. Furthermore, it is possible to conceive of feminism as not only oriented toward the "who" question in speaking but also toward the qualities of this speaking and of voice. What voices are un/heard, dis/respected, and il/legitimate? What do the voices of authority figures sound like? What voices are rendered infantile, improper, or inadequate for academic purposes?

Entangled with these guiding questions are considerations of whose voice is presupposed in the organizational, institutional, architectural, and curricular academic conditions that enable speech – and how these voice-shapings take place with each word spoken.

Voice and trans-corporeality – Voice is not just a neutral mediator

At the level of everyday, intuitive awareness, we know that the way a voice sounds affects both listeners and speakers. Tension and joy travel between, or entangle, speakers-listeners. Or, to put it another way, voice is the corpo-affective (Górska 2016), intra-active coming into being of listener and speaker. The concept of corpo-affectivity articulates the operations of corporeal materiality and affect as intra-actively constitutive and differencing. It underlines corporeal material agency and the way politics is always specifically enacted through different bodies. Górska proposes this concept to articulate how embodiment and affect are mutually constitutive in everyday life and in developing anti-discriminatory politics that work with issues of bodily and affective vulnerability (Juelskjær et al. 2020).

Caring about the corpo-affectivity of voice begs the question of how to approach this condition – when years of teaching still bare traces of years spent working to establish a voice (in many disguises – produced by the complexity of the aforementioned conditions of academic life) – in order to create a safe and stimulating teaching environment. Our voices – we thought when struggling with the legacies of women-identified scholars, intimately uncovering the feeling of not being entirely "in one's place" – are in need of some focused attention as they shift between appearing to sound too quiet, too loud, too low, too high, too dry, too uncontrolled, too slow, too fast, stuttering, lisping, muttering, strained. Somewhat paradoxically, while being managed by a competitive and accelerating academy, we believed our voices also required a bit of management. Our first experience of voice coaching allowed us to rethink this belief in reference to the question: What needs managing, and what might it mean to manage a voice?

The first coach one of us approached in Copenhagen proclaims his commitment to helping people establish "a professional voice." His website states: "You may very well be the leading expert within your field, but if you cannot communicate your knowledge using an effective professional voice, then you run the risk that no one, or very few people, will actually hear what you are saying [...] your professional voice is the voice that you should be able to switch on when you need to speak with great authority or when it is crucial that you engage, motivate, and inspire your listeners [...] when you know that you have the technique to control your voice, it will provide you with a sense of security, authority, and clout in even the most stressful situations."

For the sake of readers' comfort, it is worth repeating that to unfold the selected moments of voice coaching, we now shift to (and slip in and out of!) a first-person narrative, or a narrating "I," who is engaging in situations of voice coaching, and who is simultaneously engaged in "inner voice considerations" of the situation.

The voice coaching session one included a diagnosis of any issues when speaking and instructions in basic techniques to enhance vocal abilities. It left a vivid memory of working toward becoming a "professional speaking machine." During the session, I found it confusing, even disturbing, to listen to the coach talk and try to follow his instructions. Questions popped into my mind, like "Do I want to learn to speak like him?" and "What is it, specifically, that makes all of this feel strange?" A feminist concerned reading of the website alone could have provided some indicators to the sense of alarm and confusion felt in the situation, and may have sparked questions such as: What is "an effective professional voice" or a voice of "great authority"? How may attention to the gendering of both "professional" and "authority" play out in a coaching session? Will we debate on what sort of "authority" I may/not wish to vocalize, or how standardized that authority may be, when trained in the specific context of voice coaching sessions? Who does/not have the desire to embody the voice of control and reason? What bodies resonate with the sound of "professional voice," and how, and to what effect in situations of academic teaching and life? Perhaps the approach to voicing I encountered in the coaching session did not resonate with my desire to understand what is going on with voice in academia.

During a conversation we, the authors, had in Copenhagen in 2017, we observed that – in our cases – turning to a voice coach revealed a specific disconnect in several senses of the word:

> *A concrete example of this disconnect is a memory of my voice that was too quiet and was not able to reach the intended audience. As I was told during a group coaching session (voice coach experience three): "Your voice is located entirely in your head and it is only coming from your throat, detached/disconnected from the rest of your body." The effect is that whenever I try to speak in a louder voice, I immediately get a sore throat and cannot speak at all. The voice coach characterized my voice as "high-pitched," "polite," and "civilized." She asked: "Do you ever use your lower registers?" to which I replied: "Yes, I think so, when I swear!" And how disconnected this part is from my ("official") academic life!*

The disconnection described here also concerns the quality of a tired voice. In some cases, intensive periods of teaching can result in abusing one's voice: when you go on talking even though you feel your throat tightening and a growing disconnect between one part of you that is forcing you to speak and another that begs for mercy. So, the disconnect is also a case of not knowing how to care for one's voice and how to use one's voice in ways that do not harm oneself. The voice coach (voice coach experience three) explained that you can sense when you are using your voice in a non-harmful way as doing so is pleasurable. Therefore, having a disconnected voice also implies forgetting the pleasure of using one's voice and using it against the nap, forcing it.

Continuing on the question of disconnection, the voice coach (voice coach experience one) was working with a challenged voice ("mine") that was already somewhat disconnected, using a professional voice ("his") that confused me as it sounded and felt so weirdly controlled and cleansed. He was teaching me how to speak professionally. He was working on improving my voice by erasing the traces of whatever discomfort, insecurity, hesitation, strain that may be sensed in my unfiltered vocalization as it is/was.

The voice coach's aim, then, was to assist in installing a "professional voice" (voice of the authority) that could be switched on and off whenever needed; but oddly enough, yet another version of the disconnection became apparent. Learning how to be a "professional speaking machine" involved a process of erasure: the traces of tension and hesitation, of the effort put into speaking, of the intensities and struggles present in how voice comes to voice, all needed to be wiped out. With this particular voice coach experience, it was impossible to "stay with the trouble" (Haraway 2016) of caring for academic life and teaching, which materialized in a voice that was overly tense (or too quiet, or too trembling). There was no space to seek out possibilities to connect response-ably without necessarily detaching oneself from the materiality of a voice struggling to come to voice. Instead, it felt that the method introduced was one of disconnecting and erasing to morph the vocal system into a neutral mediator of any given content.

In contrast to this approach, the voice coach from the third voice coach experience introduced what she termed a holistic approach. "The sound of our voice," she wrote in her workshop materials, "is the last step of what is going on with a human being. [...] The voice is a very sensitive instrument: it can convey not only one's mood, but also offer an insight into our health, vitality, potentiality, and psychic condition – our successes and complexes" (Krasucka 2010: 1). This approach may lead us far away from the specificity of voice experience, but it also makes it possible to think about and with one's voice in a broader perspective. Specifically, to think with, rather than to erase, disconnections that are heard and felt and performed while coming to voice.

These experiences of engagements with different voice coaches suggest that it matters what idea or theory (regarding what enables a voice to voice) is used in voice coaching. The employed ideas and theories set the conditions for what to work with, and how, when attending to the voice.

The effort to establish "the professional voice" (as in voice coach experience one) seems to involve erasing and standardizing. It involves a specific way of controlling and mastering the voice (and academia) and a specific theory of how the sound of such a voice may engage the listener (including the one who speaks and listens to themselves while speaking). This way of coaching and voicing does not consider the specificities of which voices vocalized between which bodies can be heard as what. It does not recognize the situated and specific desires of relating in teaching that follow a particular vision of how learning and engagement are enabled. Nor does it open possibilities of questioning power relations, how they work, and how they are entangled in voice – in de/forming it, and being re/formed by it – like when you feel that you speak too loudly or too quietly, too eagerly, too passionately, too angrily. This coaching tried to remove all hesitations and irregular sounds.

But the erasure – as well as the traces of the efforts to erase – remain, as we have learned from Karen Barad's interpretation of the quantum eraser experiment: erasure is a material practice that leaves traces (2007: 313ff). When considering voicing as a labor to establish a voice, traces of that labor may be noticeable. Trembling of a voice, for instance, might be a trace of the labor that goes into erasing traces of the laboriousness, or it might be a trace of the labor put into coming to voice and a trace of the efforts and complexities that are part of this process. In academia, the "professional voice" wipes away the tensions while also being an act of conformism, hiding the institutional tensions. It is as if you are supposed to disconnect from your sense of the tensions – of working, teaching, engaging too much, of considering in situ how best to modulate the teaching-learning space using your voice – to connect with the students, which *then* becomes a version of disconnection.

The erasure of effort and labor is also visible in other areas of academic work. As Malou Juelskjær and Monika Rogowska-Stangret put it when describing the staging of research results: "All the hard work that went into enabling 'results' (grant applications, experiments, data production and analysis, developing and dismissing theories, consulting with colleagues, engaging with other researchers' results, and so forth) is made invisible, 'at the end of the day or project, the product of the physicists' activity is freed from any marks of this work" (Schrader 2012: 119). These mechanisms do not apply solely to physicists, but resonate with, and might also be used to investigate, products of the work of researchers in other academic disciplines" (Juelskjær and Rogowska-Stangret 2017: 3).

We believe that it is also possible, through voice, to make visible the work and effort that goes into the process of coming to voice in academia; to make palpable one more aspect of commonly invisible work. This is important because otherwise, one may succumb to the illusion that coming to voice is an obvious process. Appreciating the effort it takes to speak, materialized in a voice that is sometimes too quiet, too loud, too low, too high, too dry, too uncontrolled, too slow, too fast, stuttering, lisping, tense, or muttering, is an ethical gesture that invites ourselves and others to recognize this labor and its precariousness. In other words, we believe the voice allows the invisible work, effort, or trouble to be sensed because of its material, situated, trans-corporeal, and corpo-affective qualities. Norie Neumark points to this uncanny feature of voice, when she writes: "voice emerges from the body and also carries the body with it. [...] It is *haunted* by the body" (2017: 9) while it, at the same time, "*produces* bodies" (Cull cited in Neumark 2017: 28). Thus, voice is never simply a neutral mediator, sheer tool of communication, but always already entangled with bodies situated in specific power relations at play in academia. Voice might reveal them, be marginalized, or be subversive, but it cannot be easily or completely detached from them.

The trans-corporeality of voice

The hesitations, doubts, and conversations outlined above shaped the way we conceptualize the act of voicing-speaking as a multi-layered event. We understand any instance of speaking as a multi-spatial and multi-temporal affective event: speaking is a moment that is hardly a moment, as it also consists of, or is constituted by, other moments. It takes place during a period of time that is hardly linear because it embraces different past-present-future entanglements. It is a singularity, but not an individualized occasion because, in the act of speaking, in the act of using our voice, we sense how voice is a trans-corporeal phenomenon. Trans-corporeality is a term first coined by Stacy Alaimo in her writing on body and nature (2008). Applying this theoretical framework, she uncovers the ethical and political stakes of thinking bodies as "always intermeshed with the more-than-human world" (2010: 2). The concept offers potential inspiration for thinking about voice "within a broader nexus of relations between organic and synthetic bodies, actors, and forces" (Thompson and Tiainen 2017: 382).

In the context of this chapter, we see voice as trans-corporeal in the sense that it does not attest to the separation of entities and their "surroundings" – it cuts across bodies-environments, human and more-than-human, nature, and culture. Voice traverses bodies and environments; it is a sound that is both what one emits and what one hears, and is thus inherently relational (Cavarero 2005; Neumark 2017; Tiainen 2017): connecting, viscous, dynamic, cutting across various seemingly distinct entities, wave-like. Voice also goes through and across organs and body parts by moving not only vocal cords and larynxes but also other organs, including lungs, abdomens, and anuses. In the always complex event of speaking-listening, the sounds of voice vocalize how affects are circulating through, across and with bodies-surroundings. These sounds bring in different layers: personal, intimate and public, organic and technological, repressed and emancipatory or subversive – all heard, seen, and felt in how the voice comes to voice. The very act of voicing materializes and troubles different spaces, times, and power relations.

Voice is hence far from being a neutral transmitter of content (as the voice coach one seemed to hold), not only because it disrupts the private/

intimate–public binary; not only because it expresses psychosomatic states, well-being, health (aspects raised in the voice coaching session three); not only because it is influenced by a diversity of factors (the psychosomatic state of the speaking individual, the affects within/without them); but also because voicing voice is an expression par excellence of the impossibility of detachment from the above-mentioned aspects. In examining how voice comes to voice, we are able to hear the complex entanglements of how voice is produced; how various organic, biological, cultural, social, political, personal, racial, gender, class, environmental, geographical, contextual, institutional, historical, and ability-based issues are all un/voiced in a voice; how a voice becomes a voice (un/heard, un/appreciated, meaningful/less, un/recognized, not/taken into consideration); and how voicing is the enactment of matter coming to matter (world-making), as the spheres of matter and meaning are entangled in the act of vocalizing. What is more, voicing is a precarious matter. There are complex power relations at play in any occasion of voicing. Speaking enacts an entanglement of personal stories, the history of silencing minority groups (or sentences, words, ideas, methodologies, agents) within academia, ways of organizing academic debates, psychological and organic aspects, different dialects and accents, regional idiosyncrasies, race, class and gender (Arday and Mirza 2018). Voice is itself a part of the affective processes happening here-now/then-there and the processes of knowledge production, communication, and expression.

So, when engaged in reconsidering what voices do in academic situations and trying to enable a re-practicing of how to voice, longing to establish safer and more inspiring academic work environments, we find it productive to take a trans-corporeal stance instead of an individualized, cut-up understanding of voice as articulated in voice coaching session one. A trans-corporeal approach appreciates the multitudes that are in motion in the processes of voice coming to voice. Not only does it allow one to envision a voice as an indicator of what is going on in a particular individual (as mentioned by voice coach three), but also conceives of voice as a world-maker. Equipped with this theoretical take, let us return to and further unfold the speculative inquiry into (caring for) voice in academia through another moment of voice coaching. To do so, we need to begin with the concept of response-ability.

Response-ability and touch: Working on your pitch in voice coaching

Response-ability is a term introduced to grasp the complexities of responsibility in relations between human and more-than-human worlds. Haraway states: "Response […] grows with the capacity to respond, that is, responsibility. Such a capacity can be shaped only in and for multidirectional relationships, in which always more than one responsive entity is in the process of becoming. That means that human beings are not uniquely obligated to and gifted with responsibility; animals as workers in labs, animals in all their worlds, are response-able in the same sense as people are; that is, responsibility is a relationship crafted in intra-action through which entities, subjects and objects, come into being" (2008: 71).

The conceptualization of voice we are developing links directly to this understanding of response-ability. We see voice as a zone of response-ability in the sense that the way voice comes to voice enables and disables responses; it might be open, inviting to think and experiment with thinking together, and it may disconnect, intentionally or not, leaving no space for a response. A voice we fathom as response-able strives "to cultivate the capacity to respond" (Haraway 2016: 78). Drawing on Barad's (2007) use of "response-ability," Hanna Meißner claims: "Responsibility is re-imagined as an ethical injunction to work on the ability to respond to 'others,' to take care of the entanglements of our relationalities – and this implies that response-ability is tied to processes of *becoming different* in/through the response" (2014: 39). In thinking about one's voice as a zone of response-ability, it also becomes apparent that voice as response-ability entails inviting difference, allowing it to happen and reshape one's voice, thus: "*becoming different* in/through the response" (Meißner 2014: 39).

Following Haraway and Meißner, voicing is response-able in the sense that it is always already also responding to something, and in each intra-action it recreates a space, time, and an atmosphere for someone and something to respond. In a sense, then, voicing is also touching. In a literal sense, voicing touches (and is already touched by) affects, thereby moving and vibrating through one's body and the bodies of others. Moreover, touch in itself is a heterogeneous phenomenon to which Barad nods when writing: "So much happens in a touch: an infinity of others – other beings, other spaces, other times – are aroused. [.] Touch moves and affects what it effects" (2012: 206–208). Voice is charged with affect. So how to touch in a response-able, ethical manner?

The intricacy of that touch – and the ethics of touching – materializes in unforeseen ways. Such as when the voice coach instructs one of us to work on her pitch and become aware of the peaks and valleys of pitch levels, especially when ending a sentence (voice coach experience two). Let us now zoom in on this experience with the help of the "I" narrator voice.

The coach makes me read a text aloud and establishes how all sentences "end in a valley." She tells me that, when talking to students, I cannot end the sentences by going down in pitch – I need to stay on the same level or, preferably, peak (as if you are posing a question); this will keep the students energized, she tells me. Whereas when one closes on a downward movement, one "loses" or disconnects the audience and then needs to work a lot harder to reach them again when beginning the next sentence.

Since I have these sessions because my throat is constantly sore, I agree that "energizing" the students is somehow a good thing and accept that this coach might be right in her assessment. And I also trust that she is highly skilled at what she does. Nonetheless, every muscle in my body protests and

I feel an intense discomfort, bursting out: "But that must be very exhausting/aggravating/imposing/transgressive for the listener – that you keep hold of them, constantly, with your voice? When do they get a break?" It feels like the right thing to do – to let the students go, give them some (breathing, sensing, being) space! Questioning this seems claustrophobic. But the coach tells me to work with this. I read aloud and force the voice (mind-voice-muscles) to peak. I stop and protest, half laughing, half angry: "But there was a full stop; surely I must be allowed 'a valley'?" Coach: "Sure, but you will lose your audience." I: "But are they never allowed a break?" Coach: "But they do get a break; did you not have a break while I was talking? You simply get stimulated, that's all." I: "Okay, it is interesting, but..." Coach: "Try to go on, if you dare! [laughing] The important thing is that exercising, becoming aware, will enhance your flexibility, and the vitality of your voice."

Let's take a closer look at this exchange. It is somewhat striking, even paradoxical, that the coach's client is asked to practice the voicing of teaching by reading aloud from a book. This, one could argue, is not a very accurate reflection of teaching situations. At the same time, reading aloud provides a space for the voice coach to study how the client's voice "works," what its intonations and inclinations may be. It also does provoke the surfacing of the client's own apparent ethics of teaching: to create a space that invites the students in but does not grab them or keep hold of them by force. Instead, this space allows students to be present heterogeneously, to care about the possibility for the students to become different, but not in a fully premeditated and controlled manner. The client-teacher wants to engage them – but not trap them. In other words, an effort is being made to establish a space or zone of response-ability in (academic) teaching situations. How can the voice be used to honor and cultivate such a space?

In this way, the struggle of trying to reach an understanding of the reading task and the situation as such is interesting. This situation displayed specific possibilities for considering and practicing care in (academic) teaching. This ethics may not be entirely dependent on maintaining or changing a specific pitch. However, an awareness of details such as peaks in pitch draws attention to the apparently key role of affective tonality (De Freitas and Sinclair 2014: 135) of voicing in how we shape that caring. This tonality of voicing is delicate and nuanced.

Following Brian Rotman, Elisabeth de Freitas and Nathalie Sinclair adopt the concept of prosody (the ways in which an utterance is forceful through its musicality) to analyze and theorize the material and affective nature of speaking-learning: "[...] the prosodic effects of speech, [...] are the gestures within speech that lay claim to a multi-dimensional space – 'the tone, the rhythm, the variation of emphasis, the loudness, the changes of pitch, the mode of attack, discontinuities, repetitions, gaps and elisions, and the never absent play of musicality of utterance that makes human song possible' (Rotman 2008: 3)" (De Freitas and Sinclair 2014: 129–130). "[...] prosody" – de Freitas and Sinclair continue – "is an unfolding event that resists being packaged into stable units [...] the prosodic dimension of speech, on the other hand, is a continuously changing quality or differential that will forever outstrip any attempt to be unitized" (133).

This conceptualization of speaking/voicing may enable some further insight into how the voice touches, and how that touch is also a call for response-ability for the other within/out; that is, a call for the speaker and others to become and to become other, different yet entangled, in the act of voicing. We wonder how a voice coaching session (or several sessions) would unfold, if it took as its point of departure an "honoring" and caring for the prosodic effects of speech, perhaps opening a speculative space of considering the possible effects (in teaching) of the many nuances (tone, rhythm, variation of emphasis, loudness, changes of pitch, and so on) that go on in vocalizing. De Freitas and Sinclair's conceptualization of prosody also encourages engagement in the possible multiplicity of becoming different with pitch – in combination with various kinds of tone color and loudness; not least to invite students to a range of ways of engagement, energizing, and contemplation.

Voicing as establishing a voice or voices when engaging with students may then be understood as intricate work of creating a safe and inspiring space of engagement. Let us next give a little more consideration to the issue of safety.

Voicing as touching the not-yet

In Warsaw, the singer and voice coach Joanna Halszka Sokołowska explains that using one's voice is (also) about allowing oneself not to know in each and every second how the next tune will sound. Sokołowska said: "'I don't know' in every second" (Conversation 2019). This enables you to let go of how you think you should voice. It involves the risk of being out of tune or missing a tone, which is why the singer talks about how she works to feel safe or cozy in those moments – in order to feel joy and pleasure.

Bringing these considerations and practices to the realm of academia, you may come to think of how academia is seldom a safe space for voicing. Paradoxically, we find that letting go of "...knowing in every second" is pivotal to thinking. It is also pivotal when engaging in answering questions and comments from students: if you only switch on the voice that already knows it all, there is a risk of disconnecting from the vibrant activity of thinking. Furthermore, students may get a sense of the lecturer's answers as "dead matter" and be discouraged from joining in with a potentially lively activity of thinking, sensing, and learning. In this sense, controlling the voice may hurt what we hold dear about teaching and what we, institutionally, promise students.

In our understanding, telling yourself "'I don't know' in every second" is also about being committed to voice as a response-able touch. It is about letting others in with their responses, about allowing those responses to reshape how we become different, and about "acknowledge[ing] the '*unknowability* of the other'" (Puig de la Bellacasa 2017: 117). Thus, there is an ethics at the very core of voice coming to voice.

"Making room" for the other and allowing "not to know in every second" form an ethics of response-ability, in which, in addition to the above-mentioned nuances of the concept – the knowing subject is decentered, and the unknown, unforeseen, unruly voices might come to voice. There are important attendant questions to be asked, such as: how to stay with that unknown multitude of voices? How to attune to them? To which voices is it safe and inspiring to tune in, and with which voices would it be harmful to blend? What ears may be judgmental ears, and how are voices also sometimes shaped in the uncertainty of possible judgments? All possible others fold into the "not knowing," and voicing thereby remains precarious.

Now, when we approach our (provisional) conclusions, we see that the questions we posed in this chapter need to remain open in each and every academic situation of voicing, while keeping in mind the touching, haptic potentialities of voice, its response-able possibilities and trans-corporeality. The unknown with which the voice is in touch shifts how we relate to ourselves and others. In the world-and-self-making gestures of voicing, we are indeed "*becoming different* in/through the response" (Meißner 2014: 39).

Care as speculation and onto-politics

This chapter has engaged in thinking about living (precarious) academic lives through specific experiences with voice coaching. Voice coaching is – among other things – a situation one enters into with a longing to get a sense of what is going on with one's voice when voicing in teaching situations. Speaking with the voice of "a teacher" is always already saturated with past and future encounters that you and others have had in academic teaching situations. There is always already a lot going on, co-shaping the peaks and valleys of the pitch levels and the prosody – entangling and constituting the bodies of the vocal encounters. Voice coaching is furthermore often entered with a longing to relate differently to oneself and one's students in future teaching situations by vocalizing – and thereby being – differently. It implies caring about what teaching may, and may not, involve and enable.

The specific experiences from voice coaching sessions presented in this chapter made it possible for us to think about how voice comes to voice and zoom in on what is going on in that process: what values and facts are entangled in acts of vocalization. By asserting that the ways in which voice comes to voice in academic situations, such as teaching, are world-making practices, we try to stay for a while with what some scholars might take for granted or render invisible: voicing, coming to voice, using one's voice. Thinking with specific experiences from voice coaching sessions, we learned to perceive voice as touch in many senses of the term. The (ethical) impossibility of comprehending voice as a neutral mediator showed how voice is a powerful event and milieu that connects and disconnects, touching across bodies-environments, past-present-future, and there-here. Thinking about voice through the concept of response-ability positioned us/voice in touch with the other: with the responses of others that reshape how we are and become different and become plurally present, multiply entangled. Finally, voice as touching the not-yet, the "'I don't know' in every second," enabled us to be in touch with the creativity, liveliness, and openness of thinking, which is crucial for (academic) teaching. It is our hope that these insights – and the practice of noticing such micro-moments of "voice-work" – may also inspire other ways of becoming with one's voice in different settings, whether such noticing goes on systematically (as in a voice coaching situation) or is a more subtle form of heightened attention to one's everyday practice of voicing-teaching.

"We can see without being seen, but can we touch without being touched?" asks Puig de la Bellacasa (2017: 97). Along similar lines, we wonder whether conceiving of one's voice as touching also means that when voicing we are at the same time touching and being touched. In that sense, this chapter has been engaged in considering the multiple conditions and effects, or onto-politics, of voice as touch. More questions follow from this approach. Inspired by Puig de la Bellacasa's way of thinking, are we always *in touch*, in connection or relation with the other? Don't we sometimes experience voicing as a "detached touch" (Puig de la Bellacasa 2017: 99): a touch that leaves us with a feeling of disconnection or damaged relation? What about conditions that produce too much "in-touchness"? After all, "touch is not a harmless affection" (Puig de la Bellacasa 2017: 107). Making visible the dual move of touching and being touched, and at the same time, thinking about the conditions of "*being in touch*" (Puig de la Bellacasa 2017: 99) and zooming in on the complexities of this quality is a matter of thinking about how we do things. How are "values and facts […] cooked together as part of one brew?" (Juelskjær and Schwennesen 2012: 16). What kinds of worlds are we enacting in such reciprocal relationships of touch? And what do we lose when we see voice as an untouching neutral mediator? "[…] [W]hat kind of touching is produced when we are unaware of the needs and desires of that what/whom we are reaching for?" (Puig de la Bellacasa 2017: 99). Can we not ask these questions time and again as academics in search of ways to be responsible for that which we love (cf. Barad 2016) in precarious times? Can we not experiment with being in touch, asking when reciprocal touching and being touched turn into being in touch?

We re-pose these questions as we – alongside other new materialist voice scholars – consider voicing as the enactment of matter coming to matter (world-making) since the spheres of matter and meaning are always entangled in the act of vocalizing. Voicing is a precarious matter. Relatedly, re-posing the above questions is a way of gesturing toward ethics in the practices of voicing and coming to voice in neoliberal academia. By unpacking and staying with chosen moments from three voice coaching sessions and a related conversation, we wanted to reflect on the following: how to touch response-ably with one's voice? How can one ethically dis/connect with others using one's voice? What is the quality of voice-touch or touching voice that we wish to foreground and encourage? Through acknowledging the intensity and density of any given moment when voice comes to voice in academia, its entanglements with bodies and worlds, its production of selves, bodies and worlds, its plurality and the inherent uncertainty of "not knowing in every second" (which is already a multitude of spatio-temporalities), we propose an account and ethics of voice coming to voice that pays attention to intricate, ephemeral yet always present moments of ongoing labor which is put to the very acts of coming to (trans-corporeal) voice. We also point to a sphere of invisible care work that is being enacted in the efforts and struggles of voice coming to voice in academia, revealing layers of precarity deeply entangled with how our voices, bodies, and worlds matter and become different in academia.

Notes

1 We understand neoliberalism as a dynamic formation of power relations that are at play in late capitalism in 21st century, shaping relations, institutions, and subjects. Neoliberalism affects many aspects of organization of social life – among them – academia, that is, a structure of formal higher education and development of science and research at universities and other scientific institutions. This take on neoliberalism stems from Michel Foucault's reflection on biopolitics and (neo) liberal governmentality (2009, 2010), where the new political rationality emerges: market-oriented, competitive, capitalizing on the human subject, and fathoming subject as capital. Researchers today point to consequences of this approach for academia and beyond, e.g., to precarization of academic work and organizing the research and teaching processes in a business-like, highly productive fashion (e.g., Gill 2009; Standing 2011).
2 This chapter was written prior to the COVID-19 pandemic and does not include situations of online teaching or seminars.
3 Additional details concerning the sessions are placed in the end of the chapter following the reference list.

References

Ahmed, Sara. (2013). Making Feminist Points. Feministkilljoys [blog entry, 11th September, 2013]. https://feministkilljoys.com/2013/09/11/making-feminist-points/
Alaimo, Stacy. (2008). *Trans-Corporeal Feminisms and the Ethical Space of Nature*, in: *Material Feminisms*, ed. Stacy Alaimo, Susan Hekman, Bloomington: Indiana University Press, 237–264.
Alaimo, Stacy. (2010). *Bodily Natures: Science, Environment, and the Material Self.* Bloomington and Indianapolis: Indiana University Press.
Arday, Jason and Heidi Safia Mirza, eds. (2018). *Dismantling Race in Higher Education: Racism, Whiteness and Decolonising the Academy.* Cham: Palgrave Macmillan.
Barad, Karen. (2016). *Troubling Time/s: Undoing and Re-membering the Future*, The Future Lecture Series, Aarhus University. https://youtu.be/dBnOJioYNHU?feature=shared
Barad, Karen. (2012). On Touching: The Inhuman that Therefore I Am, "differences," A Journal of Feminist Cultural Studies 23(3): 206–223.
Barad, Karen. (2007). *Meeting the Universe Halfway: Quantum Physics and the Entanglement of Matter and Meaning.* Durham, London: Duke University Press.

Beard, Mary. (2017). *Women and Power. A Manifesto*. London: Profile Books.
Cavarero, Adriana. (2005). *For More than One Voice. Toward a Philosophy of Vocal Expression*, transl. Paul A. Kottman. Stanford: Stanford University Press.
Colebrook, Claire. (2017). *We Have Always Been Post-Anthropocene: The Anthropocene Counterfactual*, in: *Anthropocene Feminism*, ed. Richard Grusin, Minneapolis, London: University of Minnesota Press, 1–20.
De Freitas, Elisabeth and Nathalie Sinclair. (2014). *Mathematics and the Body: Material Entanglements in the Classroom*. New York: Cambridge University Press.
Fast, Heidi and Milla Tiainen. (2018). *Voice*, in: *Almanac*, ed. David Gauthier, Sam Skinner, http://newmaterialism.eu/almanac/v/voice.html
Foucault, Michel. (2009). *Security, Territory, Population. Lectures at the Collège de France. 1977-1978*, transl. Graham Burchell, Hampshire, New York: Palgrave Macmillan.
Foucault, Michel. (2010). *The Birth of Biopolitics. Lectures at the Collège de France. 1978-1979*, transl. Graham Burchell, New York: Palgrave Macmillan.
Gill, Rosalind. (2009). *Breaking the Silence: The Hidden Injuries of Neo-liberal Academia*, in: *Secrecy and Silence in the Research Process: Feminist Reflections*, ed. Ryan Flood, Rosalind Gill. London: Routledge.
Górska, Magdalena. (2016). *Breathing Matters: Feminist Intersectional Politics of Vulnerability*. Linköping: Linköping University Electronic Press.
Haraway, Donna. (2016). *Staying with the Trouble. Making Kin in the Chthulucene*. Durham, London: Duke University Press.
Haraway, Donna. (2008). *When Species Meet*. Minneapolis and London: Minnesota University Press.
Juelskjær, Malou, Plauborg Helle and Adrian Stine (2020). *Dialogues on Agential Realism -Engaging in Worldings*. London: Routledge.
Juelskjær, Malou and Monika Rogowska-Stangret. (2017). *A Pace of Our Own? Becoming through Speeds and Slows – Investigating Living through Temporal Ontologies of the University*, Feminist Encounters: A Journal of Critical Studies in Culture and Politics 1(1), Article No: 06. http://www.lectitopublishing.nl/viewpdf/WYQD6AIJ
Juelskjær, Malou and Nete Schwennesen. (2012). *Intra-active Entanglements – An Interview with Karen Barad*, Kvinder, Køn & Forskning 1–2: 10–23.
Krasucka, Aldona. (2010). Trener grupowy – warsztat głosowy, podstawy samodzielnej pracy nad głosem. Materiały dla studentów (teaching materials for students).
Leppänen, Taru. (2018). *Always More than Two: Vibrations, the Foetus, and the Pregnant Person in Childbirth Singing Practices*, NORA – Nordic Journal of Feminist and Gender Research 26(2): 99–111.
Mazzei, Lisa A. and Alecia Y. Jackson. (2017). *Voice in the Agentic Assemblage*, Educational Philosophy and Theory 49(11): 1090–1098.
Meißner, Hanna. (2014). *Politics as Encounter and Response-ability. Learning to Converse with Enigmatic Others*, Artnodes 14: 35–41.
Moisala, Pirkko, Leppänen Taru, Tiainen Milla and Väätäinen Hanna. eds. (2017). *Musical Encounters with Deleuze and Guattari*. New York, London: Bloomsbury.
Neumark, Norie. (2017). *Voicetracks. Attuning to Voice in Media and the Arts*. Cambridge, MA, London: The MIT Press.
Puig de la Bellacasa, María. (2017). *Matters of Care: Speculative Ethics in More than Human Worlds*. Minneapolis, London: University of Minnesota Press.
Puig de la Bellacasa, María. (2011). *Matters of Care in Technoscience: Assembling Neglected Things*, Social Studies of Science 41(1): 85–106.
Rogowska-Stangret, Monika and Olga Cielemęcka. (2020). *Vulnerable Academic Performances. Dialogue on Matters of Voice and Silence in Academia*, Matter: Journal of New Materialist Research 1(1): 23–51.
Rotman, Brian. (2008). *Becoming Beside Ourselves: The Alphabet, Ghosts, and Distributed Human Being*. Durham, London: Duke University Press

Schrader, Astrid. (2012). *Haunted Measurements: Demonic Work and Time in Experimentation*, Differences: A Journal of Feminist Cultural Studies 23(3): 119–160.
Standing, Guy. (2011). *The Precariat. The New Dangerous Class*. London, Oxford, New York, New Delhi, Sydney: Bloomsbury.
Stengers, Isabelle and Vinciane Despret. (2014). *Women Who Make a Fuss: The Unfaithful Daughters of Virginia Woolf*, transl. April Knutson. Minneapolis: Univocal Publishing LLC.
Thompson, Marie and Tiainen Milla. (2017). *Laughing Matter*, in: *Gender: Laughter. Macmillan Interdisciplinary Handbooks*, ed. Bettina Papenburg. Farmington Hills, MI: Macmillan Reference USA, 377–391.
Tiainen, Milla. (2017). *Sonic Technoecology: Voice and Non-Anthropocentric Survival in The Algae Opera*, Australian Feminist Studies 32(94): 359–376.
Tiainen, Milla. (2013). *Revisiting the voice in media and as medium: New materialist propositions*, "Necsus" Autumn. https://necsus-ejms.org/revisiting-the-voice-in-media-and-as-medium-new-materialist-propositions/?fbclid=IwAR0wqgCg6lZBmlHkzfxTdZw7AQKJ6Ct7_1PoLM1RO_S0OziUmH6NsQ8x_6A.
Todd, Zoe. (2016). *An Indigenous Feminist's Take on the Ontological Turn: 'Ontology' Is Just Another Word For Colonialism*, Journal of Historical Sociology 29(1): 4–22.
Woolf, Virginia. (1938). *Three Guineas*. London: Hogarth Press.

Empirical material

Conversation between Joanna Halszka Sokołowska and Monika Rogowska-Stangret on 11th April, 2019, Warsaw.
Voice coaching session 1: individual session (one hour), Copenhagen, (male voice coach), November, 2017.
Voice coaching session 2: individual session (one hour), Copenhagen, (female voice coach), four sessions, Spring 2018.
Voice coaching session 3: group voice coach session (2-day workshop) with Aldona Krasucka as part of the postgraduate studies on group training in Laboratorium Psychoedukacji and University of Social Sciences and Humanities in Warsaw, 2010.

Part III

Voicing the more-than-human

6 Motor-mouthing

Julieanna Preston

> … keep on … trying … not knowing what … what she was trying … what to try … whole body like gone … just the mouth … like maddened … so on … keep – … what? (Beckett 1984 216).

So reads an excerpt from Samuel Beckett's *Not I*, a play that iterates his pre-occupation with existential crisis and the improbable escape from life's pain and suffering (Westendorp 1992: 108).

> The play ends essentially where it began; movement that is, in fact, stasis. The Mouth will never cease searching for a salvation that will never materialize; the circle by definition, defies completion.
>
> (Zeifman 1976: 39)

> The script for *Not I* is known to be a nearly unlearnable text
>
> … with its exhaustively tricksy repetition and countless interjections to be spoken at such speed … simply let the words play their music out on my whole being … to speak so fast you can't think and think so fast you can't speak.
>
> (Dwan 2014)

Though only 15 minutes long, the performance makes excruciating physical demands on the primary actress. While blindfolded, she is strapped in a harness. Her face is blackened by grease and charcoal. Her mouth, outlined with red lipstick, is all that she bears to the audience through a small hole in the theatre curtain. She must speak every syllable distinctly and incessantly (Westendorp 1992: 108–109; Dwan 2014). Actress Billie Whitelaw spoke of the first time she performed the play: 'I felt I had no body; I could not relate to where I was; and going at that speed, I was becoming very dizzy … tumbling into space … falling … falling …' (Knowlson 1996: 528).

In *Not I*, the actress must develop techniques of concentration and relaxation to exercise the words, their volume, elocution, and punctuation in a tight

DOI: 10.4324/9781003390992-10

nervous rhythm. Reading the words aloud or relying on ordinary recall is not enough to affect the turmoil and tensions between the inner and outer worlds that the play aims to produce. Directed only by Beckett's short instruction 'Don't act,' it is a performance grounded in the physical activity of 'muscle memory,' a narrative that is an act of telling, not the told story (Kalb 1989: 234; Tubridy 2018: 100). Actress Lisa Dwan describes her 2005 performance of *Not I* as the 'ultimate motormouth,' a reference to a person who talks too fast and too much; someone who won't wait for a break in the conversation; they just talk over everyone else (Dwan 2013). She does not merely recite the scripted text but lives it:

> ... buzzing ... so-called ... in the ears ... though of course actually ... not in the ears at all ...,' '... mouth on fire ... stream of words ... in her ear ... practically in her ear ... not catching the half ... not the quarter ... imagine! ... no idea what she's saying ... and can't stop ... no stopping it ... she who but a moment before ... but a moment! ... could not make a sound ... no sound of any kind ... now can't stop! ... imagine! ... can't stop the stream ....
>
> (Zeifman 1976: 37–38)

Her mouth becomes the Mouth. Emeritus Professor of English at York University and scholar on contemporary British drama Hersh Zeifman describes:

> ... a disembodied human mouth, seemingly suspended eight feet in the air, trapped in the harsh glare of a spotlight ... simply a mouth, chattering compulsively ... to tell us a story ... becoming insensible ... her body, that disreputable 'machine'....
>
> (Zeifman 1976 228–230)

My interest in drawing attention to the Mouth of *Not I* is less to offer a reflection on the play or a critical appraisal of its script or any of its performances than it is to notice various ways that the play arouses heightened emotional states directly. Scholar Hersh Zeifman writes about its 1976 production: 'We are being, in effect, viscerally assaulted, we are experiencing what the Mouth is experiencing, and the compassion we feel for her is equally a compassion we feel for ourselves' (1976: 44). The command to not act as a role-playing performer but to bring the performer and the audience to the brink of mental and corporeal collapse, an unfilled or unresolved affective state affirms one's vitality. It was, in fact, Zeifman's article titled 'Being and Non-being' that compelled me to wonder if the methods producing *Not I* were transferable to my site-situated live art performances informed by New Materialist principles. Acknowledging the (mis)interpretation that my creativity wages towards *Not I* and the significant scholarship on it, I wield the trans-disciplinary tendencies of my practice to borrow liberally from the play's methods – the circle as an ideal geometry resistant to completeness and highly subject to deformation, the Mouth as a sounding organ/machine and gestures of empathy so central to live art performance.

... inhaling through the nose ...

This chapter recalls *RPM hums*, a live art performance where voice made the life force in electric motors palpable. In *RPM hums*, I acted with empathy towards other material things, specifically, mechanical fan motors. I sought to absorb, sound, repeat, mimic and emulate their vibrations; to quiver, tremble, shudder, shiver and spasm as a complementary and counter-humming material body. Proving that the motors were alive or exhibited agency independent of human beings was not the goal; the performance merely entertained those propositions as the foundation for prompting mutual and relational encounters between material 'tuning' bodies.

... inhaling through the nose ... To try ... to try to ... what?

As an essay, 'Motor-mouthing' holds fast to Montaigne's practice of essay writing, whereby a wide variety of styles, voices, methods and narratives sourced from scientific thought and subjective experience are employed to challenge fixed ideas, wonder about fundamental questions of life and human frailty and learn about one's thoughts and actions (Delbecke 2018: 5). As the essay unapologetically indulges in the pleasure of reading in a seemingly loose structure, it adheres to an emergent pattern of thought as if, in the act of writing, the idea is taking form. It takes to heart Performance Studies scholar Jasper Delbecke's statement: 'From its position between science, philosophy and art, the essay resists the reductionist reflexes towards totality and truth. In its response, the essay wants to reassess self-reflexivity, the fragmentary, the polemic, the subjective experience and discontinuity' (2018: 5).

As an essay about a performance, it is also performative in that it invests in writing as a temporally infused iterative act in synch with the rotations of a turning motor coupled with the cyclical rhythm of breath that fuels the voice. For this reason, this essay is critical writing, which performance theorist Theron Schmidt declares '... is not just about an event, but is itself an encounter between writer and work, between reader and text' (Schmidt 2018: 37). Schmidt refers to Adrian Heathfield: 'This writing is not simply upon a subject or about it but, rather, is 'of' it in the sense that it issues from it, is subject to its force and conditions ... emerges from an imperative in the event' (Heathfield 2006: 179 quoted in Schmidt 2018: 38). The performance *RPM hums*' recursive character is what prompted me to conceptualise this essay as an event of refrains. Delivered sixteen times over the course of four days, the performance repeated with a slight variation of inflection, pitch, and pace, not to mention duration, audience numbers and degrees of participation. Each delivery of the performance drew from the last; each performance generated new understanding, which was then folded into the next performance, and hence, the performance, and the essay, are processual.

> ... inhaling through the nose ... To try ... to try to ... what? ... filling the lungs ...

RPM hums was performed 1–4 March 2018 in the carpark under Te Papa, the national museum of Aotearoa/New Zealand. The work was part of the large urban festival Performance Arcade (2018). This creative work was a collaboration between spatial designer Joshua Lewis and I as performing artist; our roles are best described as 'recording body' and 'sounding body.' The performance drew attention to urban noise pollution produced by electric motors and the range of sounds they emit during prolonged periods of use.

Each performance consisted of us leading members of the public from the waterfront to the museum a short walk away. The enclosed, dark and dank car park environment is a space where visitors leave their vehicles but more so, the underworld of a building coveted for archiving and displaying New Zealand's cultural artefacts. The smooth concrete floor of the car park manages vehicular and pedestrian circulation. The ceiling, a virtual underbelly of ducts, pipes, vents, cables, fans, grills, sensors and condensing units, bears out light, water and air support systems that spatially link the car park to the storage facilities located above.

In *RPM hums*, I sounded a complimentary pitch to the motors' life force as an ethical acousmatic encounter. Following Jane Bennett's lead, I tried to take these ordinary other than human things that 'we possess and use, and are possessed and used by,' very seriously (Bennett and Loenhart 2011). Many other voices are drawn into this discussion, most significantly, Steven Connor's prosaic rumination on humming and other vocalisations (2014) and Laura Levin's concept of performing ground (2014). Paying due respect to the performance's Aotearoa/New Zealand context, I seek counsel from Cassandra Barnett's essay on Kaupapa Māori's figuration of taonga relative to New Materialism's Western philosophical approach to agency and animism (Barnett 2017). All of these sources lend to an inquiry about motor-mouthing as an empathetic, ethical gesture of enchantment.

> ... inhaling through the nose ... To try ... to try to ... what? ... filling the lungs ... holding, poised, idling ...

The palette of sounds I made were all phenomes: indistinguishable elements of speech reaching out to a magical world of noise that signifies the unintended or accidental, the crossover between meaningful and meaningless, what literary scholar Steven Connor calls the 'noisance' of the voice, or more so, the mouth-body (Connor 2014: 10–13). While crediting this organ as the producer of sounds as well as being produced by sounds, he gives the physiological mouth responsibility for ingesting food and ejecting air or vomit and the imaginary mouth idealised conceptions of 'height, lowness, frontality, backwards, softness, hardness, tension, and lassitude ...' (2014: 13). The mouth-body reveals a phantasmal dream theatre of a 'crucible in which sounds are not merely sounded, but also shaped and palpated' (2014: 15). Here Connor posits the voice between substance and form, the body's second life, lived and imagined as its subject. 'The voice is so saturated by the anxious dream of our 'life,' because it is itself one of the most important components of that will-to-life ...' (2014: 17).

My mouth-body exerted hum-like murmurs made by the delicate compression of lips intermittently sucked into the ellipse of an air-kiss, a continuous, uninflected tone in the shape of a circle with no closure, never completing but always repeating back to itself (Connor 2014: 74). They were sounds of not speaking that idled between all-vowel words and the frontier of silence. I felt the tone clenched by my pelvic floor, announced by my lungs, and tuned by my nose. I expelled the tone towards the source of the motor's whine and witnessed them meet and then merge. Unlike the motor that demonstrated a potential to persist in rotational movement forever, perhaps defying mortality, I had to take breaths; I could not hum for longer than one breath; I could not breathe in while breathing out to deliver a continuous, uninterrupted hum. I wondered if human breath modelled an oscillation more than a rotation, a mechanical system more akin to an electrical switch or valve than a constant flow over time. My organic body was pitted against the motor's mechanical body where my desire to get close, to become intimate with the motor's sound, assumed that mirroring and mimicking was the key to an audible relational encounter. The limitations of my human body to become a motor, to become like a motor, soon directed my breath to accept the complementary nature of our aural intercourse. From there on, I found skill setting a tone free very slowly such that it could join in the motor's revolutions, not necessarily in harmony, but always in gaited concert. Despite this effort, the hums were indistinct and lacked a clear location; they seemed to waft out into the space uniformly and return to me from every direction (2014: 91). These efforts lured me into a state of enchantment:

> ... with the feeling of being simultaneously fascinated and unnerved in the presence of something truly wild and generative ... an energetic current produced by the encounter between two sets of active materialities, one set congealed into a "self" and one into what is often called the "objects of experience" ... make their presence known to us or, one could say, make "calls" to which we are continually responding. The literary-visual-artistic ethical task at hand is, I think, to try to describe just what this "call" is without resorting to the framework of pre-given subjects and objects.
>
> (Bennett and Loenhart 2011)

> ... inhaling through the nose ... To try ... to try to ... what? ... filling the lungs ... holding, poised, idling ... moist ...

Director Jesse Kanda described his experience of producing *Mouth Mantra* for Björk:

> The fuschia insides of a pulsating mouth are staring back at me, sharp, shadowy white teeth warping and rotating with the surrounding flesh. Specks of saliva are dotted around, glistening with a fierce sheen. Watching the inside of a mouth can feel unsettling, nauseating, and confusing ....
>
> (Mitchell 2016)

> ... inhaling through the nose ... To try ... to try to ... what? ... filling the lungs ... holding, poised, idling ... moist ... hum what?

Humming is also often associated with the buzz of mechanical motors servicing urban interiors. Motors work due to rotational movement created by electricity and magnets. Two battery leads are connected to two springy metal pieces called brushes that allow electricity to flow in a specific direction into an electromagnet, which is made of an armature (or axel) wrapped in a coil of copper wire. Copper's high electrical conductivity makes it ideal for the efficient transfer of energy. Field magnets are positioned around the armature. As electricity is passed to the electromagnet, the field magnets attract and repel the electromagnet, which causes the wire 'to jump' and the axle to spin (Tech-FAQ n.d.).

'RPM' is an abbreviation for rotations or revolutions per minute and measures the speed the motor's armature will turn a full cycle at its rated horsepower and voltage. Under normal conditions, a typical air conditioning unit motor runs between 1,800 rpm and 3,600 rpm (Air Conditioning Systems n.d.). The design of motors is heavily reliant on Euclid's comprehensive, five-point axiomatic system, which describes geometric objects in terms of points and lines, an approach that has shaped building construction and land surveying practices significantly for centuries (Ball 1960: 53–60). In the case of motor geometry, Euclid describes a circle with any centre and a constant radial distance. Though ultimately abstract, the constructive application of this postulate is played out when drawing a circle with the aid of a pin, a piece of string and a pencil. In the simple lived gesture, the circle becomes eccentric when the pin (held at the centre) expands or contracts due to thermal gain or loss, when the string (spanning the pin and pencil) stretches or frays, or when the graphite in the pencil (at the end of the string held taut) becomes dull or held at variable angles (1960: 53–60). Such a simple exercise reminds us of how abstract geometry is and how in its material application, it deviates from Euclid's belief that his axioms were self-evident truths about physical reality. Here, we find hint of the performativity of everyday seemingly inert earthly matter.

Figure 6.1 A conducting coil of copper.
Source: Julieanna Preston (2017).

Imagine the heat and vibrations generated by thousands of rotations, the misalignment of brushes, the distortion and imbalance of magnets, the expansion/contraction of copper wire or the out-of-phase resonance of fans. These vibrations stress the ideal geometry of rotating parts; what was a circle is now an ellipse. The eccentricity is accumulative and equally vulnerable to air-borne particles such as salt, water, sand and soot. It is no wonder that the motor changes its tune from soft purring to low-pitched moaning and high-pitched whining under such duress. To remedy these sounds, electrical engineers advise to 'critically dampen' the noise using energy-absorbing materials, trap sound using acoustic enclosure systems, or apply an 'anti-noise' approach in which noise is converted to electric signals, synthesised by a processor and played back into the environment as a cancelling waveform (Noise and Vibration Worldwide Editorial Staff 2014: 11–12; Polymer Technologies 2016).

> … inhaling through the noise … To try … to try to … what? … filling the lungs … holding, poised, idling … spreading moist heat evenly through the soft ductile metal body …

Motors of air conditioning and condensing units are a primary contributor to urban noise pollution. A city dweller will attest to the low-grade and often high-pitched humming sounds that pervade their auditory world. Connor confirms the veracity of these drones and how they fatigue us by drilling into our attention and awareness (Connor 2014: 93). Their monotony is a virtue only when we wish for fresh air or need a sleep aid to escape our own thoughts. And yet, this noise persists in the background as an interruptive, irritating and aggravating ambience, often indiscernible, beyond human audibility, pitched at the infra-thin limit of there and not there, unnerving. Since performing *RPM hum*, I imagine these vibrations lurking indiscriminately, stirring up trouble for urban dwellers, unsettling their unaware material bodies.

According to social theorist Jacques Attali, sound exerts violent power over social order. He calls out anti-noise laws, policies and technological devices as attempts to keep at bay aberrant sounds, seemingly abnormal, incoherent sounds of strange others – voices of alterity (Attali 1985: 8). Sound theorist Brandon LaBelle explicates:

> … the stranger, the overheard, the foreign, and the background – these are figurations and spatialities whose relational intensities rupture and extend processes of listening and of being heard – that *unhome* us – giving promise to what we may discover from the unfamiliar.
>
> (LaBelle 2018: 72)

The disorder and dissonance of urban noise may, by virtue of its very intrusiveness, create a space open to imagining and practising new modalities of difference. In this way, *RPM hums* hunted for urban noise in the same manner as Jane Bennett and Alexander Livingston walked searching for garbage in Baltimore – noise as treasure rather than trash, noise as an affective-aesthetic exercise exposing a wildness and sensuousness of hearing beyond its triviality and utility towards its non-instrumentality (2012: 12).

> … inhaling through the noise … To try … to try to … what? … falling into the lungs … holding, poised, idling … spreading heat through the moist body evenly … red energy flushing the vascular system …

Figure 6.2 Urban wall of air conditioner units.

Source: Image by REUTERS/Vivek Prakash (https://www.pinterest.cl/pin/534872893242844649/?autologin=true).

A strand of copper wire is wound around itself as a flat coiled circular plate. Placed on a phonograph, the stylus lowered onto the surface, the needle finds its groove and its voice by way of the revolutions. Smoke wafts as the plate and needle converse in noisy intercourse (Stacey 2015).

> … inhaling through the noise … To try … to try to … what? … falling into the lungs … holding, poised, idling … spreading heat through the body evenly … red energy flushing the power system … current flow disrupted … diverted … stuck … gulp …

As a trace mineral, copper plays an essential role in forming red blood cells and maintaining healthy bones, blood vessels, nerves and immune function and iron absorption. Associated with neuron signalling, '[c]opper is like a brake or dimmer switch, one for each nerve cell' (Wilson, Medical News Today 2017).

> … inhaling through the noise … trying … to try to … hear what? … falling into the lungs … holding, poised, idling … spreading heat through the body evenly … red energy flashing through the power system … flow disrupted … diverted … stuck … gulp …

A strand of copper wire winds around my left forearm as a mnemonic device to the material's own properties and behaviour. It sweats and deforms with the rising temperature of my generous flesh, which, in turn, swells in a shade of deep pink, blushing strangulation, the hiatus before an inhale turns to an exhale.

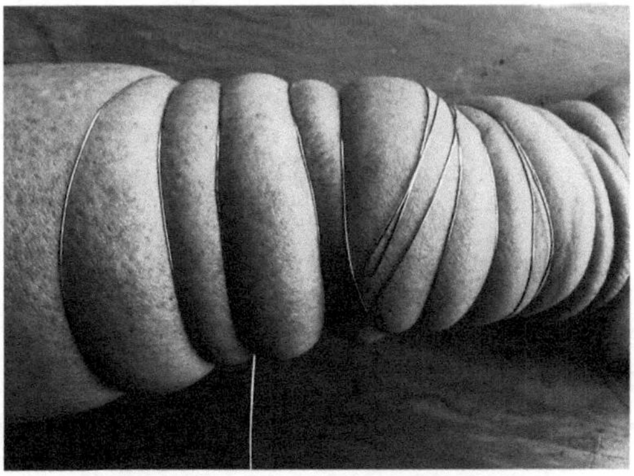

Figure 6.3 Arm and copper wire.

Source: Julieanna Preston (2017).

> ... inhaling noise ... trying ... to try to ... hear what? ... failing ... the lungs ... holding, poised, idling restlessly... absorbing heat through the body unevenly ... red energy flashing through the energy system ... flow disrupted ... diverted ... stuck ... gulp ... black out ...

In the book *Vibrant Matter: A Political Ecology of Things*, a pivotal text in New Materialism discourse, political theorist Jane Bennett asks us to abandon human-centric binary worldviews that separate the world into inanimate matter (things) and animated life (humans) (2010). Bennett promotes more responsible, ethical human engagement with our world as a means of ensuring a self-sustaining world. This view depends on understanding the world as a complex 'vibrant' assemblage in which material bodies, human and otherwise, are in a continuous state of indeterminant relational flux. It also is reliant on regarding materials and material things not as inert, mute matter but, instead, forces with unique agentic potentialities, interactions, capacities, and trajectories.

In Bennett's vibrant world, material interactions are relational, ecological and contingent assemblage-forming and assemblage-disrupting. Unlike chance in a mechanical perspective of the world, material things have an inherent vibrancy she calls thing-power, the agency of material objects to act and impact other materials, thereby producing effects in the world. 'Thing-power,' says Bennett, is 'the curious ability of inanimate things to animate, to act, to produce effects dramatic and subtle' (Bennett 2010: 6). However vibrant, a thing never acts alone; nonhuman things exist and act only in relationship with other bodies and forces, some of which are human. Vital materialism sees humans and nonhumans not as antagonists or mutually exclusive categories but instead as fundamentally collaborative, cooperative or co-constitutive, to such a degree that their differences become blurred, collapse or reform together. Often escaping human perception and understanding, Bennett recommends experiencing it with 'a certain anticipatory readiness' (2010: 5).

During *RPM hums*, it was just me in communion with motors, Joshua lurking in the shadows, and sometimes an audience lingering nearby. Because of the building's ventilation system, the museum's open hours, the tendency for a motor's metal parts to become misaligned over time and the continuous flow of electric current, the ecology of the encounter was complex and indeterminant. I was prepared to listen with my full body and being, with the promise to perceive something very familiar, always present, but rarely attenuated – tones of the motor exerting thing-power at a scale I could fathom, one motor at a time.

What happens in the practice of searching for material vibrancy when the scale of concern is increased exponentially? What happens when the distributive agency of a national electrical power grid is not taken seriously? In 2003 over fifty million people across North America lost power in a blackout that involved a networked assemblage of many motors, bodies, and materials distributed over a huge terrain. Bennett's analysis of the event exposed the tragicomedy of how government policy at national and state level, along with corporate capitalist venture decision-making, enabled the black-out. More important to this essay, she tracks the strange behaviour of electrical current seeking a path of least resistance (literally) as 'a cascade – of voltage collapses, self-protective withdrawals from the grid, and human decisions and omissions. The grid includes various valves and circuit breakers that disconnect parts from the assemblage whenever they are threatened by excessive heat' (Bennett 2010: 25). The systemic rupture of the grid's hot current made electricity's typically invisible agency, highly visible.

Bennett cites a newspaper article that described the blackout as the moment the power 'grid's heart fluttered, a clear example of a human tendency to anthropomorphise and personify non-human objects and occurrences attributed to human wilfulness and sense of privilege over the rest of the world' (Bennett 2010: 25). To give the power grid a heart which can flutter makes it easier to grapple with as something strange and foreign; the power grid is made metaphorically animate. Bennett dispels any misunderstanding that the theory of vibrant matter resurrects or asserts new notions of animism; it does not attribute a soul to plants, objects or natural phenomena nor assert the presence of supernatural power or being to create or organise the material world (Bennett 2010: 48).

> ... exhaling ... noisy ... trying ... to try to ... hear now? ... felling out ... the lungs ... hollow ... less ... even expenditure ... warm ... gut wrenching ... invisible ... gaseous force ... squeezed through ... flow depleted ... across the soft cords ... black hole ... pitch ...

RPM hums was performed on the other side of the world from where Bennett lives. The world I live in, the place I call home, is a bi-cultural country. The fact that I am both an immigrant and Pākāha (white person) influences how I engage with Māori knowledge, protocol and artefacts. Neither token, superficial nor casual, this interface is part of my everyday life. It takes ongoing time and commitment to become familiar and versed with patterns of thinking, living and making, framed by a very different worldview than the one my Western European/American upbringing instilled. The relationship I have with this place is as a life-long learner. The lessons I learn are not made by circular rote repetition but as spirals, augers digging down into the ground and leaves spinning in an upward whirlwind simultaneously. There are moments as this learner that I have (mistakenly) tried to take a shortcut by equating New Materialist vitality, thing-power, to 'mauri,' the Māori relationship between people and all living things, simply translated as 'life force' (Hēnare 1988). There are other moments, such as *RPM hums*, where my experience of life is led by my understanding of Māori concepts (limited as it might be) such that the concepts are tangible.

I performed *RPM hums* in the underworld of a building coveted for archiving and displaying New Zealand's taonga, loosely translated as valuables or treasured things (Barnett 2017: 24). I imagined that the vents provided me direct communication, perhaps even ingestion or inhalation of the dust of these precious material bodies. It is here that air-borne particles of the cultural treasures get swept up, sucked up, deposited, spewed and liberated from the protective security of the museum through a maze of air ducts. Perhaps the motors served as some kind of New Materialist translator/transducer, motors whining and whinging while pushing the particles of culturally heavy stuff through a hole from inside to outside? Was this related to the Mouth in *Not I*? Opening wide to let the air of a motor's fan fill my mouth, leaning my shoulder into the vibrations of a condensing unit, putting my ear against the rumbling sheet metal and puckering my lips to shape a like-tone were gestures well beyond the mechanics of hearing or listening; it was what musician/educator Christopher Small called musicking, which:

> ... establishes in the place where it is happening a set of relationships, and it is in those relationships that the meaning of the act lies ... relationships between person and person, between individual and society, between humanity and the natural world and even perhaps the supernatural world.
>
> (Small 1998: 13)

Figure 6.4 Still image from performance, day 3.

Source: Joshua Lewis (2018).

RPM hums was as close as I have come to truly experiencing material vibrancy. And yet I wondered, was this 'mauri'?

As a writer who seeks and creates immanent relations between contemporary molecular philosophies of art, postcolonial and decolonising thought, and indigenous (especially African) and Māori aesthetics, Cassandra Barnett (Raukawa, Ngāti Pākehā) guides me to see the differences between this Western theory and this indigenous concept. In her essay *Te Tuna-Whiri: The Knot of Eels*, Barnett writes that 'taonga can be considered animate and alive because they instantiate ancestral hau (life breath), mauri (life force) and mana (spiritual power) in the present. Hau, mauri and mana, and hence taonga, have enduring force and efficacy—they may amend the course of things here and now in unpredictable ways' (Barnett 2017: 24). Her words open a space for me (and other non-Māori) to understand the degree of my (our) otherness. Writing with care, patience and grace for the principles and deep thinkers that proceed her, she reminds us that such terms as 'animism' and 'spirit' underpinned by metaphysical dualisms are not found in te ao Māori (the Māori universe) (2017: 41).

> Such proximities and interleavings afford closer enquiry into where recent Western animisms (as enabled by new materialisms and philosophies of the posthuman and anthropocene) and indigenous world views meet—and where they part ways. Like Māori taonga, contemporary art can invoke a cosmic vibrant materialism, an interconnectedness of all things, and a concern for the role/responsibility of the human within this. But somewhere around the assigning of 'anthropomorphic' personalities and behaviours (and even names) to things, the Western philosophies still tend to become troubled. Taonga Māori land us in a place where 'animism' has profound efficacies, yet does not exist as a critique-able term or concept.
>
> (2017: 25–26)

Barnett points to curator Amiria Henare's acknowledgement that taonga have the capacity to produce and reproduce relations as live subject-less objects that need conscious performative engagement, mostly importantly, through the korero (stories), the taonga, as the specific efficacy of ancestral form, tell of their identity between spirit and thing or objectness (Barnett 2017: 29). In *RPM hums*, there was an almost imperceptible orchestration of vibrations, the retelling of stories over and over again, something not immediately visible yet felt, heard, performative and affective all the same.

Conscious of Barnett's heed to regard the function of taonga 'beyond the bounded horizon of precolonial Māori life and beneath the surface of other more visible or dominant currents, systems and dynamics—including global, capitalistic, corporate and aesthetic ones' (Barnett 2017: 24–25), I tread lightly in her footsteps to embrace and practise, in the best way I know how, the notion of making things 'more visible ... a matter of hau ... calling us to attention … calling forth … insist upon our attention, our custody, our care of them …' (2017: 25). Reading Barnett's text again and again, I am reminded of an unsettling feeling I experienced during the performance. With eyes closed, my body oscillated with the vibrations of the motors. At peace with my own opacity and the opacity of the taonga (and motors), I came to know that I do not know (2017: 39–40). While my humming vocalisation initially drew memories from my youth singing softly to stuffed animals, forest birds and geckos and, much later, to my baby daughter, the mouthful, the bellyful. Such memories of air laced with the hau, mauri and mana of taonga prompted an urge to plug up the vent, close my mouth and rush to the interior of the museum to return, give back, to regurgitate, what I had collected, what wasn't mine, and, my onus to honour.

> … hailing … nose to nose … trying … to try to … be now … failing … the long white cloud … hallowed … more … warm … more gut wrenchingly … visible … force … squeezed through … pitch black … across the ocean … loving to be loved …

Sci-fi writer J. G. Ballard tells a story about sweeping up the sonic residue of everyday urban life – a 'frenzied hypermanic babel of jostling horns, shrilling tyres, plunging brakes and engines' (Ballard 1960: 106). The story is set in the future when all existing forms of music have been replaced with ultrasonic music, which is perceived on a subliminal level beyond the normal human auditory range. Though silent, it is thought to produce a sensorium of richer emotions, textures and compositions. Mangon, a mute boy with hyper-acute hearing, vacuums all the leftover traces of sound from the surfaces of architectural spaces using a device called a 'sonovac.' Each night Mangon deposits the sounds he has collected in '[a] place of strange echoes and festering silences, overhung by a gloomy miasma of a million compacted sounds … remote and haunted, the graveyard of countless private labels' (1960: 122). When his love is rejected by an ageing opera singer trying in vain to restart her career with a live public performance, he retaliates by using the sonovac to drown her voice.

> The voice exploded in his brain, flooding every nexus of cells with its violence … grotesque … an insane parody … [r]ough and cracked … jerked … sudden precipices of gasping silence … plunged through the volcanic torrent ….
>
> (1960: 154)

Then, in complete despair, he destroys the sonovac with his own bare hands, letting the utterly painful noise of her voice loose to pollute the opera house, the village and the world at large.

Ballard's short story wrestles with sound as a material substance and quantifiable yet qualitative auditory matter. In his world, the sounds of everyday conversation and phenomena are imperfect. It begs me to think beyond Western European notions of noise as unwanted, destructive, disordered, toxic and dirty (Pickering and Rice 2017: 27). Hugh Pickering and Tom Rice link sound and dirt by way of the phrase made famous by Mary Douglas: 'Dirt is matter-out-of-place' (Douglas 1966: 44). They cite her:

> ... there is no such thing as absolute dirt ... no single item is dirty apart from [i.e. outside of] a particular system of classification in which it does not fit ... Where there is dirt there is system ... dirt is not an independent, objective attribute of something ... but ... a residual category [of things] rejected from our normal scheme of classifications ... a label for all events which blur, smudge, contradict or otherwise confuse accepted classifications.
>
> (Pickering and Rice citing Douglas 2017: 6)

For Pickering and Rice, noise is not like dirt; it is dirt. Noise is sound that is ambiguous and anomalous; while transgressing a system of commonly respected values, it announces a political difference. For Bennett, trash, another form of dirt, is stuff to ignore but, in new materialist terms, also stuff that calls attention to itself and its existence in excess of human constructions, meaning and activities (Bennett 2010: 4). To have a dirty mouth is one thing; to hold dirt in one's mouth is another. In Douglas' terms, to amplify vocally, as in *RPM hums,* is to exaggerate in a deliberate creative act that resists the normative order, something sticky or viscous, unstable, and blurring the boundary of the self and the other in an intense and entrancing pleasure (Pickering and Rice 2017: 10–11). The bits of fibres and granules of sand that blew into my mouth through the vents were dirt, bodies, mouths, motors, holes, taonga in the making – nothing shy of political, especially in Aotearoa where, as Barnett taught me, matter is always *of* place.

> ... hurling ... ear to nose ... trying ... to try to ... be heard ... faltering ... the dirty white mouth ... more ... thick ... more visibly thick ... pungent force ... thrown across the pitch ... longing to be loved ... of ... what?

During *RPM hums*, Joshua tracked me searching for motor noises. We often found ourselves in a game of 'tag' amongst the columns and cars, a game that turned into 'marco-polo' when my visible body diminished to only an audible humming voice haunting his ears with sounds so close that he said he could hear my saliva accruing and my heart beating through his headphones. It was so extremely unnerving to him to be so intimate, in my mouth aurally but way over there, out of sight and proximally distant. He responded by pointing the camera towards the ceiling to faithfully follow the ducts and pipes as they were in relation to where and how I was. This shifted the performance from actions of hunting and capturing akin to Mangon's act of collecting in Ballard's story, to gestures of 'performing ground.'

> ... hurling ... ear to ear ... trying ... to try to ... disappear ... fostering ... the dirt stuck between white teeth ... more ... invisibly moist, thick

and pungent ... care ... stretched over the cords ... longing to be loved ... of ... what? ... inhaling through the nose ...

Professor Laura Levin introduces her theory of 'performing ground' by asking what it means when site-specific performances turn their attention to ecological resonances of environment (Levin 2014: 110). While tracing the impact of live performance shifting from the theatre to urban everyday spaces, Levin reveals how in the historically constructed, patriarchal, figure/ground, male/female, culture/nature binary equation, women and environment have become synonymous with background material. Levin expands common notions of camouflage relative to blending in, hiding, imitating, mimicking and mirroring in order to describe art practices in which the body is presented as coextensive with its material surroundings (Levin 2014: 6). Unlike other feminist theorists who might consider blending into the environment as merely self-preserving subterfuge, debasing, submissive, assimilating, yielding to mastery, erasure or retreating to the shadows in a diminutive manner, Levin resurrects the act of camouflage as one of power, choice, care, and devotion to excess, zealous repetition, and physical contact capable of producing 'a different morphological relation between self and world: the apprehension of bodily boundaries as porous, the self in sensuous relation (a body "among millions of other celestial bodies")' (2014: 121). Levin calls out this form of mimesis as a threshold, a hyper-immersive spatiality, an ethical and political accounting, an interlocutor, a limitless resource, a sign of adeptness and adaptability, and an evocative form of empathy while cohabiting with the external world (2014: 14). To mimic, to emulate, is a complementary sign of respect, a gesture denoting empathy.

Figure 6.5 *RPM hums* video. See also Performance Video link at Routledge support material website: https://www.routledge.com//9781032480022

Source: Joshua Lewis (2019, https://vimeo.com/319627282).

I can only hope that *RPM hums* performed ground and, in doing so, advanced Levin's notion of 'the environmental unconscious,' which seeks to interact with all the elements of a site and situation as alive and in play, effectively supporting Bennett's project to reject how 'the image of dead or thoroughly instrumentalised matter feeds human hubris ... preventing us from detecting (seeing, hearing, smelling, tasting, feeling) a fuller range of nonhuman powers circulating around and within human bodies' (Bennett, 2010, ix). *RPM hums* inserts sonic space and experience into Levin's theory, which is heavily geared towards the view, the picture, seeing and visibility. *RPM hums* considers sound (inclusive of noise) as an all-immersive medium, forces flowing in and amongst everything, a pure transfer of energy, 'swirls of matter, energy and incipience' a notion that differentiates it from philosopher Timothy Morton's concept of the 'hyperobject,' which casts things as holistic objects and nonlocal entities manifest in vast, unknowable, spatial dimensions (Bennett 2012: 227, 239–232). Following Bennett's lead:

> I find myself living in a world populated by materially diverse, lively bodies. In this materialism, things-what is special about them given their sensuous specificity, their particular material configuration, and their distinctive, idiosyncratic history-matter a lot. But so do the eccentric assemblages that they form. Earthly bodies, of various but always finite durations, affect and are affected by one another. And they form noisy systems or temporary working assemblages which are, as much as any individuated thing, loci of affection and allure. These (sometimes stubborn and voracious but never closed or sovereign) systems enact real change. They give rise to new configurations, individuations, patterns of leanings, and affections. Networks of things display differential degrees of creativity, for good or for ill from the human point of view.
>
> (2012: 231)

In the video that emerged from the performance, the female figure is left behind, or perhaps she has become so opaque she has been absorbed by the environment. This is not unlike the artist Francis Woodman, who Levin discusses at length; Woodman attempts to immerse herself into the architecture, the wall and its linings (Levin 2014: 50–52). The visual aspect of the performance video of *RPM Hums* offers no sign of my mouth-body except for the reverberations that linger in my humming (roaring, squeaking, rusty, soft and crooning) voice that crawls, slides, turns and pivots with respect to the ceiling apparatus and its spatial tentacles to the museum's taonga.

According to Small, the living world is tied together by 'the news of difference,' a phrase borrowed from anthropologist Gregory Bateson referring to the static difference between figure and ground as it is scanned or changes dynamically over time (Small 1998: 54). The gestural language of biological communication 'is seamless, no vocabulary, no units of meaning, seamless, without gaps, and depends not on quantity and the amassing of discrete units such as words or numbers but on shapes, forms and textures – patterns, in fact – patterns of relationships' (1998: 59). I would contend that *RPM hums* employs an older form of gesture called paralanguage where 'para' refers to something subsidiary, something to one side or altered, even improper or wrong, something devalued according to today's highly verbal society (1998: 61).

The literalness of the gesture is to stand of place beside something else, in parity, *with* something else, such as an out-of-tune motor. Two eccentric bodies following, collaborating and co-existing even in inharmonious moments. It is to hum like a motor, a motor-mouth rendering violence to the status quo of paradigms negating material vitality, not behind a curtain and through a hole as a disembodied mouth, but with the enchantment of difference heard in the currency of the motor's surging, out-of-round, copper-coiled, magnetism.

References

Air Conditioning Systems. n.d. *Air Conditioner Motors*. Accessed 6 June 2020. https://www.airconditioning-systems.com/air-conditioner-motors.html.

Attali, J. 1985. *Noise: The Political Economy of Music*. Translated by Brian Massumi. Minneapolis/London: University of Minnesota Press.

Ball, W.W. Rouse. 1960. *A Short Account of the History of Mathematics*. New York: Dover Publications.

Ballard, J. G. 2001. The Sound-Sweep [1960] in *The Complete Short Stories*. Flamingo: London: 2001.

Barnett, C. 2017. Te Tuna-Whiri: The Knot of Eels, in *Animism in Art and Performance*, edited by Chris Braddock, Australia: Palgrave Macmillan: 23–44.

Beckett, S. 1984. *Collected Shorter Plays*. New York: Grove Press: 216.

Bennett, J. 2010. *Vibrant Matter: A Political Ecology of Things*. Durham and London: Duke University Press.

Bennett, J. 2012. Systems and Things: A Response to Graham Harman and Timothy Morton. *Literary History* 43 (2): 225–233.

Bennett, J. and Livingston, A. 2012. Philosophy in the Wild: Listening to 'Things'. *Scapegoat: Architecture/Landscape/Political Economy* 2: 12. Accessed 13 January 2022. http://www.scapegoatjournal.org/docs/02/02_Bennett_Livingston_PhilosophyInThe.pdf.

Bennett, J. and Loenhart, K. 2011. Vibrant Matter, Zero Landscape [interview with Jane Bennett]. Accessed 6 June 2020. https://www.eurozine.com/vibrant-matter-zero-landscape/.

Connor, S. 2014. *Beyond Words: Sobs, Hums, Stutters and Other Vocalisations*. London: Reaktion Books.

Delbecke, J. 2018. Tracing the Essay in Contemporary Performing Arts. *Writing & Performance, Performance Research* 23 (2): 5–12.

Douglas, M. 1966. *Purity and Danger: An Analysis of the Concepts of Pollution and Taboo*. London: Routledge.

Dwan, L. 2014. *Samuel Beckett's Not I: Lisa Dwan mouths off -- video*. https://www.youtube.com/watch?v=e4qaRp4t4jg [Available 16 June 2020].

Heathfield, A. 2006. Writing of the Event, in *A Performance Cosmology: Testimony from the Future, Evidence of the Past*, edited by Judie Christie, Richard Gough and Daniel Watt. London: Routledge: 179–182.

Hēnare, J. (Tā Hēmi Henare).1988. *Waka huia*. https://teara.govt.nz/en/video/41077/ko-te-reo-te-mauri-o-te-mana-maori [Available 7 June 2020].

Kalb, J. 1989. *Beckett in Performance*. Cambridge, New York, Melbourne, Sydney, Port Chester: Cambridge University Press.

Knowlson, J. 1996. *Damned to Fame: The Life of Samuel Beckett*. New York: Simon & Schuster.

LaBelle, B. 2018. *Sonic Energy: Sound and Emergent Forms of Resistance*. London: Goldsmiths Press.

Levin, L. 2014. *Performing Ground: Space, Camouflage, and the Art of Blending in*. New York: Palgrave Macmillan.

Mitchell, A. 2016. *Behind the scenes of Björk's 360-degree 'Mouth Mantra' video* (originally published in Dazed). https://www.dazeddigital.com/music/article/32826/1/behind-the-scenes-of-bjork-s-360-degree-mouth-mantra-video [Available 16 June 2020].

Noise and Vibration Worldwide Editorial Staff. 2018. *Reducing Noise from Fans*. 49(1): 11–13.

Performancc Arcade. 2018. Curated by Sam Trubridge, Wellington Waterfront, New Zealand.

Pickering, H. and Rice, T. 2017. Noise as 'sound Out of Place': Investigating the Links between Mary Douglas' Work on Dirt and Sound Studies Research. *Journal of Sonic Studies* 14. https://www.researchcatalogue.net/view/374514/374515/0/500

Polymer Technologies. 2016. *Damping Techniques for Noise and Vibration Control*. https://blog.polytechinc.com/damping-techniques-for-noise-and-vibration-control [Available 6 June 2020].

Schmidt, T. 2018. How We Talk about the Work Is the Work. *Writing & Performance, Performance Research* 23 (2): 37–48.

Small, C. 1998. *Musicking: The Meanings of Performing and Listening*. Hanover: University Press of New England.

Stacey, I. 2015. *Copper sounds*. [Online]. https://vimeo.com/130025609 [Available 5 June 2020].

Tech-FAQ. n.d. *How do Electric Motors Work*. [Online]. https://www.tech-faq.com/how-do-electric-motors-work.html [Available 6 June 2020].

Tubridy, D. 2018. *Samuel Beckett and the Language of Subjectivity*. Cambridge, New York, Melbourne, Sydney, Port Chester: Cambridge University Press.

Westendorp, T. 1992. Catharsis in Beckett's Late Drama: A New Model of Transaction? *Samuel Beckett Today/Aujourd'hui* 1: Samuel Beckett 1970-1989 (1992): 106–113.

Wilson, D. R. 2017. Health benefits and risks of copper. *Medical News Today*. https://www.medicalnewstoday.com/articles/288165 [Available 6 June 2020].

Zeifman, Hersh. 1976. Being and Non-Being: Samuel Beckett's Not I. *Modern Drama* 19 (1): 35–46.

7 The non-standard performance with the singing theremin

Tero Nauha

Introduction: The theremin and non-standard performance

The Soviet scientist Leon Theremin (Lev Sergeyevich Termen) invented the theremin in 1919. As one of the first electronic instruments, the theremin uses a relatively simple technology, featuring a frequency or pitch antenna and an amplitude or volume antenna. It's a space-control instrument, meaning that you play it without touching either the volume or the pitch antenna. Stephen Montague explains how the theremin operates on the principle of heterodyning: two signals of sound, which have nearly the same frequency with each other, cancel out the original signal to produce a third frequency. The theremin uses two radio frequency oscillations, both above the human range of hearing. Yet the resulting third frequency is humanly audible, within the 15,000–18,000 HZ range (Montague 1991: 20).

The theremin is an exceptionally liminal instrument that evokes an "emotional technology," a term Hans-Joachim Braun (2009: 148) applies to the *technostalgia* that has arisen in recent decades – a revived interest in transistor radios, vinyl records, console games, and other outdated devices. Often, the theremin represents not an avant-garde ideal but a lingering 19th-century romanticism as a counterpoint to that ideal. Certainly, the revival of the theremin has become more prominent with the rise of digital technologies like MIDI and sampling. At the same time, the actual playing of the theremin suggests mysticism and a heightened sense both of the virtuosity of the instrument and of the virtuoso who operates it. José van Dijck notes how "media technologies and objects are often deployed as metaphors, expressing a cultural desire for personal memory to function like an archive or like a storage facility for lived experience" (Dijck 2009: 111). Perhaps because of this cultural desire, technostalgia is mainly linked to popular music or sound art instead of the field of "classical" contemporary music.

When we consider the playing of the theremin, however, as an example of performance art, it goes beyond the virtuosic or the magical and becomes aligned with a grotesque liminality. Considered from a performance viewpoint, the use of the theremin is spatially and temporally *in-between* and *entangled*. It offers not only romantic technostalgia, but also a rite-like

DOI: 10.4324/9781003390992-11

performance of transformation or transition. Performance is essentially ambiguous, slipping around classifications or standards, betwixt and between positions and postures. The liminal is the place of the "unwritten laws of the gods," writes Audronė Zukauskaitė (2010: 69). A technological object gives form to something that otherwise would have been felt only through the senses. A liminal object can also be viewed through its indeterminacy of energy, and this is especially the case for the theremin as an instrument you play without touching. Performing the theremin proposes something invisible or undetectable, an energy to be inferred but not experienced in any direct sense – something non-standard or metaphorical.

These aspects of liminality and attendant uncanny inhumanity were already noted in the early performances by one of the first theremin virtuosos, Clara Rockmore. There was a specific interest in her playing's machinic quality – its robotic and alien affect – and even today the quality lingers, flirting with the technostalgia for this instrument. Thereminists generate uncanny, mystical, and demonic effects, creating sounds resembling sirens from outer space. This mesmerizing quality has the aural consistency of fog mingled with the bursting of uncontrollable circuits, as if the musician descends on a chimeric tumult of material, immaterial, and metaphysical alliances. A musician who's able to contain these diverse forces seems to have seized control of supernatural powers.

This chapter barely touches upon the technical side of the theremin, or on the life of Leon Theremin as a Soviet engineer in the West. Instead, the focus is on the affectivity of playing the instrument, and on the use of the theremin as a way to create a stance, a pose – a posture of the human voice.

I understand affect here in the wake of Baruch Spinoza's philosophy, or as Brian Massumi (1998: 60) argues, affect is constitutional to social stratum and the variable of systems. In *Schizoproduction* (2016), I consider how, in performance, there is a collective stratum or consistency of affect that sews together the performers, the audience, and technical devices through collective speech, which, in turn, refers to Félix Guattari's (1984: 43) concept of "*agencement collectif d'énonciation*" – an ensemble of subjectivation. However, I do not take affects as pure intensities, but also as a capacity that can produce aversions or passions, and therefore, it is closer to Spinoza's vital philosophy of power, endeavor, and *conatus*. Affect is the capacity to be in relation to the milieu. Gilbert Simondon writes how the process of individuation is always coupled with a milieu, always understood with its milieu, and mediated by affect (Shaviro 2006). The individual must have the affective capacity to establish any relationship and, in Simondon's terms, to undergo a process of individuation, or in other words, to exist in processes of becoming: "grasping the genesis of individuals within the operation of individuation as it is unfolding," writes Muriel Combes (2013: 3). Affects appear through milieu.

This means my interest in the theremin is neither in virtuosity nor in the taming of the shrew, the savage machine. Instead, I am interested in the affective trembling or vibrational co-creation that the theremin offers – what I call *thereminING*. I'm interested in the singing voice of the theremin, less a mimicry of a human voice than a posture of a nonhuman voice, a voice we hadn't heard before the theremin's invention. This is a processual intra-action – indeterminate, indefinite, but not entirely uncanny, *Unheimilich* but also *Heimlich*. It calls not for nostalgia but for an inhabiting "home-affect," with a reverberating compassion in that homely strangeness (Ettinger 2009: 2). Playing the theremin is an *intra-action*, not an event of two separate things coming together, player and instrument, but an internal process of differentiation, where the difference is the phenomenon. An interaction presumes prior existence of independent entities, but, following Karen Barad's (2003: 815) formulation, "it is through specific agential intra-actions that the boundaries and properties of the 'components' of phenomena become determinate and that particular embodied concepts become meaningful" in their never-completed actualization. Thereminining is such a process of intra-action, a "mutual constitution of entangled agencies" (Barad 2007: 33), where the technical object – the theremin – has a possibility for a mutual response.

As noted, you play the theremin without touching it, and you can use any part of your body for this – or any other conductive matter. Though this is a process of quilting or plaiting entanglement, it's also a process of diffractive interference. The action involves bending, not overlap or blending, with the elements reading diffractively through one another in the act of theremin playing. Imagine a vacuum filled with leptons and bosons, or other fundamental particles. In that vacuum, no straight line or reflection exists, but in every instance, deflection excludes difference and involves a zigzagging diffraction. As physicists Brian Cox and Jeff Forshaw describe such a vacuum, the diffractive impact is like walking toward a bar in a crowded pub (Cox and Forshaw 2012: 2017). During the walk, "one gets buffeted from side-to-side and ends up taking a zig-zag path towards the bar," so that in effect the "empty space is not empty" (Haraway 2004: 70).

Another invention by Theremin, the terpsitone, is operated by the movement of a dancer around space-controlling antennae, and as with the theremin, all the movements need to be precise and minute. When thereminining, you try not to swing your body, because the movement might bring your head closer to the metal rods than your hands are, and therefore cause the instrument to play out of tune. The left hand regulates the volume, but it can also create crescendo, diminuendo, accent, and arpeggio. If you do not use your left hand, you hear continuous sound. At the same time, the proximity of your hands to the vertical rod on the right side can create glissando, tremolo, or vibrato.

In this chapter, I explore the "voice" of the theremin and its similitude to the human voice. As an electronic instrument, the theremin needs to be manipulated before it produces sound. The human voice is usually produced by exhaling from the lungs, and this exhaled sound will eventually stop due to the body's automatic need to inhale at some point. To vary the sounds of a theremin, however, you need to perform a deliberate act that will interrupt or change its baseline pitch. If you don't move at all, the theremin will produce the same pitch indefinitely, without variation. Any production of sound requires a decision, a motion to make it begin, move, or end. Yet compared to the voice of a living being, a theremin won't stop involuntarily – potentially, if a conductive material is near, the sound can go on forever.

Concepts for performance thinking

Some of the concepts used in this chapter require further elaboration, since they are either obscure or used in unfamiliar ways. First, the concept of *non-standard* here refers to the non-philosophy of François Laruelle. Laruelle argues that all philosophical thinking is foreclosed from the Real. As an alternative, he proposes non-standard thinking, which renounces all metaphysical mirages of the Real or of "the science of Being" (Laruelle 2013a: xxii). Laruelle's non-philosophy involves thinking *according to* the Real, reality, or the universe, as opposed to thinking *of* those subjects. Thus, non-standard thought is thinking *alongside* and *entangled with* reality. This approach is not transcendent but radically immanent: for non-standard thought, there is no superior knowledge about reality or the world. Rather, all thought is, at least theoretically, equal. All thought – and especially the superior thought of philosophy – is essentially positional.[1]

Laruelle finds the potential for a redeemed unity in the dialogue of opinions, points of view, and positions – through what he calls decisions, or *decisionality*. Decisionality divides, decides, cuts off, and reduces. It's never absolute; yet, it may appear, for instance, as an image of thought, or as a concept distinguished from a thing. A decision is the invariant of all philosophy, a split of facts from data, a split between logos and physis. A decision cuts off and withdraws into a position. Yet in connection with the concept of the agential cut, or agential realism, Barad notes: "phenomena do not merely mark the epistemological inseparability of observer and observed, or the results of measurements; rather, phenomena are the ontological inseparability of agentially intra-acting components" (Barad 2007: 33). For both Laruelle and Barard, decisionality directs us not toward the transcendental or the philosophical, but toward the "philosophical horror" in our realization that thinking has always existed.

Decisionality is, therefore, the literal and epistemological *cut* and not merely a reductive, descriptive, or analytical instrument. In applying decisional *performance thinking* to the theremin, the standards of thought are, in a sense, speculative positions for defining, reducing, and reflecting. Almost inevitably, philosophy and philosophical standards of thought are necessary for all artistic practice. Yet through the non-standard twist – through the entanglement of decisionality – we see that philosophy is only one of the materials with which performance thinking works. Surely, we need a decision now as much as it was needed in the early development of sound technology and cybernetics to distinguish signal from noise, the voice of Edison from the crackle of the tinfoil.

For Laruelle's concept of non-philosophy or non-standard thought, no mode of thought has the capacity to model the Real; yet, all thoughts are determined by the Real, by its radical immanence. In the *fictioning* sense, all modes of thought are equally valid through the generic nature of thought. Laruelle posits that philosophy believes it holds *the* mode of the Real, and further believes the mode is universal. This belief creates the *position* of each philosophy in relation to other philosophies: every position is a struggle against another position, but philosophy, as a whole, regards itself as the capital form of thought (Laruelle 2012: 261). The philosophical decision, or cut-off, is an "operation of transcendence that believes (in a naïve and hallucinatory way) in the possibility of a unitary discourse of the Real" (Laruelle 2010: xiv). The events of the world include playing an instrument like the theremin, and performance thinking about that playing. These events are real, yet in terms of decisionality, we experience them as the effects of the philosophical decision, the proto-event and the self-positing event, and the event of the "worlding" (Laruelle 2000: 178). Here, the use of philosophy is an act of auto-positioning and also a break from the previous state of philosophies, an impure act of distributing the transcendental and the empirical.

In this chapter, I will also refer to the term *posture* as roughly the opposite of a position. It is, in fact, an entangled and unclean concept because the posture is a concept-thing. Often postures are denigrated as something ridden with sticky sentimentality, silly, unauthentic, etiolated, and obscene. Postures are essentially performative and are often viewed as a generic and cheap recycling of formally superior concepts and thoughts – like the sound of the theremin. But for non-standard thinking, it's precisely this stickiness that gives the posture its power.

Both philosophy and postures are determined by radical immanence: they are both equally present alongside this reality. Postures are valid forces of thought, and because of their entanglements, postures may be even more valid than philosophy. However, performance thinking emphasizes that all modes of thought, from philosophical position to non-standard posture, are practices. A posture is not a cut but a certain kind of haziness, a mist between the thing and the conceptual representation. It's not so much the conceptual meaning that determines the "reality" of the object as the relation between the representing and the represented. A posture is not mimicry guided by conceptual misconception or slack; it is a practice and a style of thought. A posture is *in* the agential cut, or *in* the continuum of entanglement tied immanently with the conceptual-material *reality* of the object – the artwork, performance, clothing, film, or poem.

For Laruelle, a posture is corporeal and in the midst of the Real, but it is also more naïve than a decisional position, since any decisional position is always relative and exclusive of other positions, and always alienable. Postures are "confusing" because they don't follow a speculative image of thought.

They aren't figurative – rather, they are *figural* and intransitive kernels of reality (Laruelle 2013b: 42). A posture is generative and, to some extent, processual. This processual quality is essential for our performance thinking – our sense of probabilities, vacillations, and convulsions, all of which are ultimately non-binary and peripheral to positions needed for the production of knowledge. A posture is always performative and etiolated, without ontological finality, without a cut to distinguish one thing from another. Laruelle offers the example of a photograph, where the result is an indivisible whole with all of the optical, chemical, and perspectival properties entangled. The picture is itself a "photo-fiction" (Laruelle 2012: 19), a posture that is always performative but not citational.

Postures are precarious, uncanny, crimped, and entangled. They are without a standardized figure, and always anchored on the potentially feigned performance undermining any standard position. Performance thinking is relational and affective entanglement in the specific terms of a specific event, and any decision as to those terms always arrives too early or too late, hiding the entanglement behind some standard concept or universal rendering.

In addition, aside from Laruelle's non-standard thought, this chapter also refers to terminology created by the artist and psychoanalyst Bracha L. Ettinger. At first glance, Ettinger's terms at least partly contradict Laruelle's, but my aim is to show that for performance thinking, they share commonalities in rebuking binary positions. In thereminig, the focus is on the continuum of entanglement to be found in the practice of the performer working with the technological instrument. While playing the theremin, your hands, overall gestures, and body reassemble in immanence (*tempus*) the material-conceptual relationship between a living (conductive) body and the metal rods of the electronic instrument. In this context, I will apply the term *severality*.

Ettinger uses the concept of severality in connection with matrixial subjectivity. She writes that the matrix weaves woman both as a subject and as an object in the "axis of *heterogeneous severality*" (Ettinger 1997: 390), whereas the phallic plane posits woman *either* as a center (such as *jouissance* or nature) or as a lack. The matrixial is a space of diffraction and severality, shareability and hybridity. For Ettinger, this space is opened up in her practice of painting, where the act "perforates" the space. The act doesn't happen consciously or rationally, but neither does it correspond to "the unconscious as defined by the chain of repressed signifiers" (Ettinger 2001: 112).

Through my practice with the theremin, my aim is to postulate how an act of severality is taking place, via the diffraction and entanglement between a technical object and a live, conductive being. Severality provides no fixed point for scopic or social regimes; yet, it is not a "split" or Uncanny in the sense of the horror of castration, as it would be in the regime of the phallic gaze. Ettinger (2002: 234) writes how it is impossible to gather traces of a matrixial gaze: you can only join them, become interwoven and entangled in the matrixial web, the space of diffraction. Matrixial subjectivity refutes opposition and fusion. It's a shareability that evades collective community – a severality, not a collective unconscious (Ettinger 2006: 219). Severality is an assembly of several entities, differentiated yet proximate, formed through plaiting and diffraction.

This assembly is affectual and material. It takes a specific form because affects are transferred in the *borderspace of severality*, like traces transmitted and redistributed, opening new affective passageways (Ettinger 2006: 222). Severality of this kind depends on sensitivity: conducted by the senses, it attracts and repulses and resonates in an atmosphere that is phantasmatic and *jouissant*. In this sense of theremining in severality I cannot detach myself from this borderspace, yet I would agree with Ettinger that such a shared milieu is not for collectivity but a diffracted assembly, a process of individuation.

Playing with the theremin

The sound of the theremin is iconic, recognizable from Cold War Hollywood science fiction movies like *The Day the Earth Stood Still* (1951) and a whole slew of space invasion films with amusingly interchangeable titles: *The Thing From Another World* (1951), *Phantom from Space* (1953), *It Came from Outer Space* (1953), and the immortal *Plan 9 from Outer Space* (1959).[2] A sense of oddball wackiness has long been associated with the instrument, as Robert Moog recalls in his foreword to *Theremin: Ether Music and Espionage* (Moog in Glinsky 2005: ix). Through the 21st century's ongoing wave of technostalgia, the theremin has aroused new interest without ever losing its special aura of quirkiness. Indeed, as far back as the 1930s, the virtuoso Clara Rockmore had already established the image of the mysterious thereminist, when she first introduced the instrument to a broad North American audience.

Rockmore, sometimes called the "high priestess" of the theremin, was an Eastern European immigrant to the United States, like Theremin himself (Braun 2009: 148). She was a virtuoso violinist, but after meeting Theremin in New York in the 1930s, she abandoned her violin career and started practicing with the new instrument. She gave her first recital with the theremin in New York City's Town Hall in 1934. The concert left the audience spellbound, and *The New York Times*, on October 31st, praised the way that "such artistic music floated on air from a source seemingly uncontrolled by human effort" (Montague 1991: 21). Planted on the stage with her crimson dress, Rockmore stood completely still, only moving her hands. Again, since any change in the proximity between the player's body and the rods alters the pitch and the volume, a thereminist usually avoids any movements except those necessary for playing. Such immobility further emphasizes the supernatural and hypnotic effect of the theremin concert, where it seems that performers are pulling the music out of thin air.

The theremin's otherworldly presence continues to attract artists today. Tom Waits used a theremin in the musical *Alice*, composed by him and directed by Robert Wilson, with the theremin played by virtuoso Lydia Kavina.

In an interview, Waits states: "you play the air, and you're in contact with the waves – somehow [it] does something to you on a more genetic level and can heal the sick. Raise the dead. Apparently" (Phipps 2002). Kavina herself describes how a particular "insecurity" and "trembling" characterize the instrument:

> When playing the theremin there is no mechanical memory, no connection between a note and a hand or finger position ... no form of visual or tactile orientation when playing. The player's only source of orientation is her ear; she plays – in this sense, it is like singing – only intonations that are imagined by listening inwardly.
>
> (Stoll 2006)

The instrument is difficult to control: it is hard to hit the tune at its center. This is one way that the theremin is quite different from the more easily managed synthesizers or MIDI instruments. It really is a bit wacky, with its posture of a human voice simultaneously suggesting a posture of a viola. At the same time, this similarity and resemblance create an uncanny and alien affect. It's easy to define through the mimicry of the human voice – but this is precisely the view that my chapter aims in part to dismantle. The theremin has the posture of a human voice, but it is also a ventriloquist's doll, a slightly uncanny instrument. Even so, its impact on popular music and visual culture has determined much of the fifties and sixties sonic field, from B-Movies to the Beach Boys' "Good Vibrations" to Jimmy Page's use of the theremin at Led Zeppelin concerts.[3] The theremin has come to represent the sound of an unknown and maleficent threat from outer space, or of an alien noise created by a mad genius.

In my work with the theremin, I've employed a few spatial effects but have used the instrument as a controller for a granular synthesizer built by musician and researcher Derek Holzer. In this combination, the microphone has served as an input to the synthesizer, and the theremin as the controller of the synthesizing, where I split or create stuttering effects for voice and singing. Sometimes I play the instrument along with the Moog monophonic synthesizer, or with some minimal spatial effects. Whatever the case, the process always seems to involve a mess of cables connecting these contraptions and the microphone to the audio mixer, the amp, and the speakers. In a dimmed space with a bit of spotlighting, this all may look magical and mysterious. Even so, my interest in playing the theremin has been performative, focused on the posture of the vocal qualities of sound. I am not a singer, not even a sound artist, but my approach is clearly driven by the conceptual – by the entanglement with different bodies.

When the theremin is joined to a voice with granular synthesis, this creates a partially closed circuit, slightly different from a plain voice and very different from the sound of the theremin alone. In these circumstances, the theremin must be played with painstaking precision in order to reproduce the note or granule of the voice. Most of the time, while playing the theremin, I will fail to hit the note in its center. It seems that the theremin introduces, as an aspect of its sound, the impossibility of controlling it fully, especially since I lack the virtuoso training of Kavina or Rockmore. This isn't the case with the theremin sounds of Jimmy Page or the Beach Boys, but in those situations the theremin is mostly used affectively. It tends toward an alien agency: rock stars perform on the instrument to demonstrate their control over maelstrom forces, unleashing and ultimately ruling the spirit of the inhuman machine.

I'm driven by the same affectual interest in forces, but less with the goal of controlling or unleashing them than with encountering the conceptually obtuse instance of non-standard severality. I resist calling this uncanny or mysterious, because those terms obfuscate the conceptual indeterminacy of the performance. They arrive too early with a definite and universal rendering, often related to the standard concepts. It's no surprise that Kavina calls thereminists slightly mad. Seen in that light, a theremin is rendered comical and poorly responsive to the needs of a fine vocal tonality. It becomes too easy to dismiss the instrument as an obscure and evidently failing curiosity, like a dated side-show effect. What we miss, with such a dismissive approach, is the entanglement in thereminging.

Montague writes that Theremin once received a phone call from Lenin, who invited him to present the theremin – this new "Soviet" innovation – for a private demonstration. Later, Theremin told Montague that Lenin was able to play the theremin pretty well, and almost immediately. Supposedly Lenin performed the "Internationale" for Theremin. This seems highly unlikely, yet it gives flesh to the idea of the theremin as a miraculous Soviet invention – and to the way that Lenin conceived of its use (Montague 1991: 19). Lenin wasn't always keen on experimental artists, but he viewed the theremin as a useful tool for conveying a grand image-of-thought of the revolution. Clearly, in his mind, it had a political function: the invention was made for the "Internationale," much as Russolo's Intonarumori musical instruments were invented for the Italian Futurists. Since its beginnings, however, the theremin was, to Lenin's way of thinking, safe from the Intonarumori wander-lines of indeterminacy as expressed in *L'Arte dei rumori* (*The Art of Noises*, 1916). The theremin is a robot that needs to have a function. At least you could play a tune with it and produce more than mere noise! Nevertheless, while the refrain of the "Internationale" was used for reterritorializing comrades from the lines of flight emerging in the Russian Revolution of 1917, the theremin was never the instrument employed for these performances. Counter to Lenin's promise, the mass production of theremins never took place in the Soviet Union. Later, when Theremin toured the United States, he played the instrument not as the widespread and popular voice of the "Internationale" but as the unearthly voice of quirky Soviet contraptions, to put it in terms of standard thinking on standard procedures.

In contrast to the above terms, the theremin's indeterminacy and non-standard qualities are *lived* in the performance. This means they are not standard procedures. A performance as lived practice is performing what it is proposing, yet these propositions are themselves always performative. Most importantly, they don't have any predetermined causal relation but are immanently lived. To call this uncanny is to arrive too early or too late at a definite conclusion, which is always rendered universal or at least standard. When Kavina calls thereminists slightly mad, thereminists are overdetermined as uncanny, and in terms of standard thought they arrive too early or too late to the scene. This, of course, is why most use of the theremin in old movies now strikes us as comical and camp. *The Day the Earth Stood Still* is a well-made film for its time, and yet the uncanniness of its theremin music sounds just as absurd to us today as the faux-theremin music in the spectacularly incompetent *Plan 9 from Outer Space*. In overdetermining our sense of the theremin as quirky and uncanny, we are too quick to dismiss it as absurd. We thus miss what it might reveal to us beyond the bounds of standard thought, and how we might instead experience it in terms of non-standard performance.

Stop in the center of it

The sound of the theremin does not, in its unmodified basic pitch, particularly resemble either a human voice or a viola. For those resemblances to exist, you must create them. Through your actions, you may make the instrument perform a posture of a voice. Put it this way: if you tell people that your hobby is to play the theremin, they're often curious to know what it sounds like. In response, you might mimic the theremin with your own voice, producing a rapid glissando that resembles a slightly deranged police siren. The glissando ends when you no longer make the effort to sustain it, so you can return to your regular breathing. If you were then to turn on the theremin itself, the theremin might be experienced as sheer noise, because its basic pitch will sustain itself potentially forever, like the engine sounds during the early industrial period, or the humming of an urban landscape. If I let the theremin play continuously in my studio, with a relatively low volume and so that neither my body nor any other conductive material will affect the pitch, it won't be long before the noise dissolves into the background. In time, of course, even that background noise will disappear from my attention, like the sounds of the ventilation system in the office space where I'm writing this – sounds that I usually don't even notice.

The significant point, in terms of the *posture* of the theremin, is that the instrument isn't mimetic – it's a posture because you decide for the instrument to *become* mimetic. If the theremin in the room continues to play almost below hearing level, the sound is still immanent and materially present. Through your hand gestures and simultaneous mental decisions, these sounds transcend themselves and maybe even become music. They transcend through the gestures of your decisions, where you both control and stop the movements. During these gestures and decisions, the sounds of the theremin are immanent and materially present the entire time, like the electricity running through the instrument. The posture is there, as it was in the process of the teenage La Monte Young, who used to listen to the high-tension lines in Idaho and who later transformed this material entanglement of wind and electrical wires into a composition for brass instruments, "The Second Dream of the High-Tension Line Stepdown Transformer" (1962). The Idaho high-tension lines still continue their material entanglement, and so does another experimental posture in the "eternal" work *Dream House,* presented at the Tribeca studio of Young and Marian Zazeela. The *Dream House* continuously plays different frequencies of sine waves and has been doing so since the early 1960s. Breathing is a continuum, like the sound of the theremin or the high-tension wires in Idaho, but to act on these sounds – to speak, stutter, or mumble – is a performative posture and a decision.

What happens with the theremin – or with the different breathing techniques used in singing – is that we reach the so-called uncanny valley and potentially pass through it. We become aware of the materiality or the organicity entangled with the air that's needed to produce voice. We may feel that something is pulling our hands and fingers in relation to the theremin rods, as if the theremin is requesting these movements, or as if we were able to *feel* the affect of the electromagnetic fields. The theremin is uncanny in part because it offers a *posture* of exhale that can be prolonged infinitely. It's an instrument of virtual breathing, like an automaton from outer space, or a figure like the Turk, Wolfgang von Kempelen's fake mechanical chess player. My interest here is not to reach for the veridicality of the theremin or to locate the nature of its uncanniness. I'm interested in the hands and gestures that reassemble the material relationship in the space between the body and the instrument's metal rods, and how this reassembly makes the theremin sing – or, more accurately, makes the *severality* of us sing.

Human voice has long-term associations with agency and rational subjectivity, corporeality and identity. The uncanny *doppelgänger* of the machinic and wailing sound of the theremin has this semblance to the living being. The theremin doesn't resemble a *bunraku* puppet or a prosthetic hand – the examples of the uncanny valley that roboticist Masahiro Mori (2012: 98) gives in his article introducing this concept. In regard to Mori's affinity scale, the theremin resembles an industrial or toy robot: charming and enticing, and creepy only on certain occasions. Yet, I feel there is some kind of a valley in the sound of the theremin when my hands are *entangled* with the air and the instrument's sound.

Imagine Clara Rockmore, performing with her expertise. Then imagine how the theremin, as a space-control device, is also controlling the musician through agential entanglement, inviting the player to engage the instrument with complete physical and emotional poise and control. In an interview from 1945, Rockmore says: "You must not only hit a note … but you must hit the center of it. You cannot register any of your internal emotion at all" (Glinsky 2005: 156). The agential cut is what creates the uncanny notion of the human hands swaying like prosthetic arms with artificial muscles. The cut represented in the uncanny gives form to a question: where is that noise coming from? The almost robotic precision of the thereminist's hands, like the uncanny face of an AI robot, is based on the speed of affect. Mori (2012: 100) writes that, in the creation of "the dynamic sequence of facial deformations" that go into a robot's artificial smile,

> the speed of the deformations is crucial. When the speed is cut in half in an attempt to make the robot bring up a smile more slowly, instead of looking happy, its expression turns creepy .... [B]ecause of a variation in movement, something that has come to appear very close to human—like a robot, puppet, or prosthetic hand— could easily tumble down into the uncanny valley.

The posture of a theremin performance, with its peculiar stiffness and machinic attributes, turns posture into the uncanny, and an agential cut appears. We retract to the standards of philosophizing either too early or too late.

Decision to perform

In performance thinking, the posture in performance is not mimicry, since a performance resembling the original source would produce a theremin-as-if-it-were-singing. Rather, the posture is non-standard, in line with Laruelle's non-philosophy, where a posture of thought is performative and an identity of practice (Laruelle 2013: 178). A Laruelle specialist in non-philosophy, John Ó Maoilearca, writes that non-philosophy practice is thinking through *samples*. A non-standard performance is performing thinking (Laruelle and Ó Maoilearca 2015: 179). This is always uncertain and in motion – motile thinking with potentially weak samples, which take form in postures. In contrast to the secured position of proper thought, such as philosophy, a posture is performative and parasitic in terms of the Austinian speech act. In this sense, postures are infelicitous and merely citational of positions. Yet, from the perspective of the non-standard, infelicitous postures are not *lacking* anything in their entanglement with reality.

Your hands, near the theremin's antennas, change position slightly, flexing or releasing, joint by joint. There is no sense of touching anything concrete. You're only grasping at thin air, entangled in the weakest position of thought: nothing to hold on to. In order to make sense of anything and to present some concrete and standard tune, you need to have an image of thought – for instance, an invisible string – so that you may transcend or ascend within this image.

Example: in a certain arrangement, I play the theremin with a Wah pedal. My intention is to search for a movement where, using my hands and my right foot, I can make one phonetic sound or a combination of them, resulting in something like an *uh-oh*. For a simple phonetic semblance like this, I need to use the gestural skills of my whole body. If the semblance is produced, it's always parasitic and infelicitous: a performance. The postures are partial and dividual, like strips of behavior, which have no origin but only reiteration. The postures are not only parroting the original, as would be the case in the very strict sense of the Austinian speech act.[4] The postures are being performed for others and are always framed by the performer, but they're also *dividual* entanglements. The point is not to mimic the monosyllable *uh-oh*, but to test as fully as possible the entanglements with the theremin, with the performer's body, and with the possible effects of the mixing board. How can I feel the severality of the *uh-oh*, the body, and the theremin in order to modulate a *posture* while the posture is simultaneously being formed? This is the question I try to investigate through theremining.

Standard thought would argue that each thing has its place, its one ontological status. All relationships, material and immaterial, would be ordered through a relation with conceptually distinct positions: the theremin does *not* sing, but it sounds *like* singing; or performance thinking is not properly thinking, but it is *like* thinking. In other words, the original would be found through philosophical inquiry.

In that standard world, we wouldn't take postures or non-standard thinking seriously. Yet my argument rests on the non-standard postulate that all non-standard gestures or postures are performative, whether they're etiolated or not. They are performative because of their entanglement with dividual severality. In regard to Melanesian culture, Marilyn Strathern writes of dividuality and relationships:

> [P]ersons are frequently constructed as the plural and composite site of the relationships that produced them .... [T]he body is a social microcosm *to the extent* that it takes a singular form .... [I]t contains within it diverse and plural relations.
>
> (Strathern 1988: 13–15)

The dividual is distributed, and in the process of differentiation boundaries are determined. These boundaries are always performative, even in their postures.

When we perform, we know certain things *already* – because we perform from the dividual immanence, and not only as a rational and transcending ego or as an individual character. In performance, we know patterns that are motile and already present, patterns we contact via a posture, without ever creating conventional or conscious positions. We don't perform in order to resemble a virtuoso. In the absence of a "finished product," as Paolo Virno writes, the performance is always motile, relational, and occasionally banal, and only able to produce the *uh-oh*: "the purpose of action is found in action itself" (Virno 2004: 51–52). The posture is not something to be informed by any property or skill but by diffraction and entanglement.

When I practice, I don't aim to learn to become precise, but I do practice the postures – my physical stance in relation to the theremin. I practice how to deviate through *dividual relations* from the standard decisions of music. These performative movements are not made from one *position* to another, but rather involve a mastery of the probable, a mastery of where the postures might end up taking us.

Where does this form of practice lead? Surprisingly, toward compassion and a sense of the homely. There is, in fact, compassion in thereminingbeyond the uncanny. Compassion has to play a part in entanglement, beyond unnecessary cynicism and the decision it entails. In playing this way – performing with performance thinking – both subject and object are woven in severality, not in *either-or* but *both-and*, what Ettinger would call a space of diffraction and severality, shareability and hybridity. The performance isn't primarily scopic but somatic, and it is neither holistic nor made to reach a whole. This is what I understand with the matrixial gaze, which is interwoven with bodies, not necessarily with the vision. Any part may function in this diffractive space of the matrixial gaze, this assembly of severality.

The homely affect is the place for artistic practice, inseparable from the process of habituation, cohabituation, *milieu*, or co-affecting. This *Heimlich*, Ettinger writes, is possible through trans-sensorial *reverberating* compassion:

> … [T]he compassion of homely strangeness arise[s] on an aesthetic dimension which is proto-ethical … In the matrixial field of vision, but also of sound, touch, etc., awe and compassion are bound with fascinance.
>
> (Ettinger 2009: 2)

Theremining, as a practice of performance thinking, co-affects in a field of sound and touch. This happens not in-between the theremin and the player but in their reverberating cohabituation and in the friction needed for their motile postures.

The intra-action of mutual constitution is simultaneous, and for one to be able to listen to the theremin in the performance, another diffractive cohabituation is involved, a different kind of entanglement. This cohabituation may also involve vision and the scopic regime – even a scopic regime of the uncanny. These are closed systems and sequential operations that proceed simultaneously with an environment but not necessarily as compassionate cohabituation (Luhmann 1993: 35). Instead, the thereminist may be *viewed* as a *freak*. The performance is not taken seriously but converted to meet certain standards. The different elements – the movement of the hand, the electric circuits of the theremin, the materiality of air, hands, and body, the possible awe and compassion – have not only a relation *in between* things but a relation via the ontology of motility. All the concepts of posture, severality, entanglement, and diffraction can be read through a *both/and* motility: both thing and concept, but not in a conventional or stable way.

In his anthropological research on the Ifá divination practice in Cuba – where *aché* is the grace of Orula, the patron deity of Ifá – Martin Holbraad discusses the white powder full of virtues spread on Orula's divining board. It is the powder that enables the Ifá initiates to divine. *Aché* is power *and* powder, abstract and concrete, a concept *and* a thing, both/and, in a non-antinomical way (Holbraad 2007: 204). The diviner is a *babalawo*, but a *babalawo* without *aché* is not a *babalawo*, and powder that is not properly consecrated is not *aché*. The meaning lies in the inherent motility and not in the powder or the performative context. *Aché* is not a thing – it's a space-time for ontological transformations. The ontology of powder *is* power. The specks of powder are a multiplicity in which the motility of Orula speaks through the forces of participation. Severality is a concept-thing: it is powder and power in the same motile ontology of relationality.

Severality in artistic practice, too, makes the practice *nonantinomical*. We may not get there through positions of proper thinking. A thereminist without motile ontology with the theremin is not a thereminist, because the thereminist has no power without such a severality. The power of the theremin is in motility and severality; it is in the reverberating cohabituation, in Ettinger's terms. The playing of the theremin *is* the voice of the thereminist, the severality singing through the movement of the hands in the reverberating thin air. The proposition here is to discontinue to regard these severalties as uncanny resemblances of the human voice.

The affective noise of the theremin

A theremin is a perfect example of a posture, because its sound is often regarded as citational, parasitic, mimetic, or etiolated. The instrument hardly has a canon or standards, but it has an infelicitous timbre from "elsewhere."

My practice of using vocals and synthesized experimentations began first with a prepared text, which I read out loud and recorded. The recording was then edited and pressed into records, on acetate vinyl. Finally, in my performance lectures, I played the records on DJ turntables. I did scratches and slides, stuttered my speech, altered the pitch. I also mixed in a live voice reading the same text. Later on, I began to experiment with the theremin because of its timbre, its similarity to a human voice.

In contrast to the assemblage of vocal cords, bones, and muscles of a human body, another kind of materiality is processed for the voice that emerges through a vinyl record's miniscule mountain range, nearly invisible to the naked eye, but perfect for the stylus to travel. When played via another kind of signal processing, we hear the voice of the speaker through the loudspeakers. This voice speaks like a ventriloquist's dummy or an automaton. It has a different temporality than the same person speaking the same text in front of the audience. The process is, in essence, not so different from one of the first recordings ever made, where Emile Berliner recorded his voice in 1889 on a shellac record. It is also not so different from the process of Thomas Edison singing "Mary Had a Little Lamb" to be engraved on a tinfoil cylinder. Berliner's voice intra-acted with the excrement of the Indian female lac bugs, *Kerria lacca*, also known as shellac – whereas my voice intra-acted with the nitrocellulose needed for the dubplates.

Considering Edison's voice, Marie Thompson (2017: 41) remarks that

> it is not just the inventor that speaks. The "voice" of the invention is also audible. The crackling of tinfoil and the whirring of the machine underline and sometimes distract from Edison's recital.

The voice recordings in early technology did not evoke awe and excitement. Instead, as Louis Chude-Sokei (2016: 72–73) writes, people who heard the recordings reacted with

> less wonder than absolute terror. They were encountering the human voice disembodied, the human itself disembodied, and the spectacle—both aural and visual—of speech rendered no longer exclusively human.

This is certainly the imitative uncanny, but might it be noise as well – the whole assemblage of crackling and tin-foiling that created the early listeners' sense of terror? What kind of presence is that material severality, which we have already learned to distinguish both from noise and from voice? Might it have been the uncanny intra-action that evoked such horror? The posture of a human voice entangled with tinfoil and shellac, a voice that barely resembles but rather has motility and reverberating cohabitation, and which also is a mockery of a human voice, as if the recording were ridiculing us? A voice that is both material and intangible, power and powder. Such are the postural propositions, the non-standards, when encountered for the first time. Only a few decades later, and before sound came to cinema, Rockmore toured with the theremin, performing in the United States. Her repertoire consisted mostly of standard classical music, with no pieces specifically composed for the theremin. These spectacles were received with awe and excitement but also anger and horror.

In terms of phallic structure, it is particularly horror, hysteria, masquerade, or parody that connects with mimicry. In this framework, the theremin is merely a noise-machine, a gimmick act aiming to replicate the human voice or the viola. It is received through the othering act accompanied by private fantasy. Yet in terms of matrixial severality, theremining is not to be read as confrontational or submissive – it's not split and not based on phallic difference, but on diffraction and hybridity. Theremining is a diffractive textile of severality, which, for Ettinger (2004: 89), "does not mean an endless multiplicity of singular individuals, but rather a limited multiplicity – a severality – that traverses subjectivity. The severality is a necessary result of the affective shareability that underlines it." Instead of being seen as othering, such performances are to be received as homely cohabitations. In phallic structure, theremining diminishes to masquerade in order to find its position and proper concept. For severality, such reductive scanning gives way to shareability and cohabituation. A decision cuts off and transcends, while cohabituation has a certain non-standard haziness, motility, and blur – one might call it *noisy* – in the continuum of entanglement tied immanently with the concept-material reality of the object, thing, artwork, performance, clothing, film, or poem.

In order for us to hear Edison's voice, we need to reduce the amount of information so we can (barely) distinguish his speech from the noise of the tinfoil crackle. Noise is obvious but evasive, slipping in and between what Thompson (2017: 3) calls the "productive, transformative force-relation and a necessary component of material relations." Noise is decidedly noise, but only partially so, since there is no music or sound without noise, and it betrays all binaries, because noise is not *either/or* but *both/and* – it is a conjunctive synthesis. Noise is not the Dionysiac encountering the Apollonian. In affective relationality, noise occurs at the level of matter that constitutes any encounter. It is severality: lungs and ears, shellac and human muscle, air and metal rods. Thompson (2017: 11) writes that noise is often felt as well as heard. Noise abides in the relational severality of the dividuals, their materialities.

But noise is also a moralistic concept, used to determine position and supremacy. The music of slaves, for instance, has often been dismissed by those in power, the colonizers, as noise, worthless and incomprehensible, alien and intimidating, with an unwelcome promise of decay, disorder, and rupture.[5] The Soviet composer and musician were performing noise, which since has been turned into technostalgia. Yet they are stripped of their motile power, their severality, and may appear only as caricatures, like Dionysiac action-figures that may no longer constitute any revolting and transformative relation. The severality of noise and voice in theremining should not be defined through binaries, not through their resemblance or oddity, but through their affectual relations.

What does a hands/theremin severality *do*? The hands aren't so much *playing* the theremin as providing an implicit practice in severality. Nina Sun Eidsheim notes how vibration is both a metaphor and a material relationship between humans and things. These relations are always present, without a priori definition or originality. Vibrations are dynamic and processual, changing across and according to materials. Eidsheim (2015: 17) writes:

> If singing and listening both constitute the process of vibration across material, they are always present – or, more correctly, always occurring. In short, listening to, making, and manifesting music is a vibrational practice.

The sound of the theremin's voice/noise severality affects and moves us in a homely manner in the *milieu* where the uncanny is ever-present. Yet without the binary division, in the motility of this material-conceptual performance, theremining and listening are materially dependent on the multisensory register, since they are vibrational and affective. In theremining and listening to this "voice," we linger on the boundary, as in the matrixial compassionate encounter – or we encounter a derivative in-betweenness, where the theremin may not have any wrong positions. This is a severality of performing and listening, passing through different vibrational relations. Theremining is a boundary object, and Susan Leigh Star and James R. Griesemer (1989: 408) have described such objects as "simultaneously concrete and abstract, specific and general, conventionalized and customized ... [and] often internally heterogeneous." Boundary objects may maintain coherence in intersecting worlds. Therefore, I propose to view the voice of the theremin as a boundary object without a presupposed epistemological primacy of one positionality that emerges from decisions.

Clara Rockmore left the audience spellbound with her virtuosity in playing the theremin. She played as if she were pulling the astounding sound out of thin air (Rodgers 2010: 9). Her approach to the theremin created séance-like performances, where she remained absolutely still in order to control the pitch antenna, and where even the smallest trembling or gestures alluded to the inner motions of the thereminist. Always, however, the theremin itself remained an ominous and mysterious black box with protruding antennae, like a spaceship's control board or a radio transmitter to the stars. At the same time, as the contemporary thereminist Lydia Kavina has suggested, the theremin may also be viewed as an instrument expressing the vulnerable human psyche (Braun 2009: 149). And yet, fortunately enough, this music from the Ether always retains a bit of its childish and amateurish timbre: the theremin remains stubbornly resistant to standards. The theremin informs performance practice with the value of impermanent motility, where the value of amateurish affect is always present in compassionate encounters – encounters that are the reason for voicing and performing at all.

Notes

1 "Clearly, Laruelle's own critiques suggest the ongoing possibility – not of transcendent judgment but of an immanent, context-bound evaluation wherein equalizing thoughts are preferred over authoritarian ones. The Laruellian model is not one of 'anything goes'; rather, its form of democracy allows us to prefer those thinking practices that tend towards pluralism rather than authoritarianism whatever form that might take in *any given situation*" (Cull Ó Maoilearca 2017: 746) (emphasis in original).

2 The composer for *The Day the Earth Stood Still* was Bernard Herrmann, whose first film score was for *Citizen Kane* (1941) and whose final film score was the experimental horn-and-percussion music for Scorsese's *Taxi Driver* (1976). Generally considered one of the best film composers of his era, Herrmann worked on some of Hollywood's most technically accomplished movies, and his use of the theremin probably cemented the instrument's role as a mainstay of science fiction productions. At the other end of the spectrum, *Plan 9 from Outer Space* is regularly cited as the worst movie ever made, and its technical ineptitude at all levels

is legendary. Even its theremin sounds are a cut-rate cheat: instead of an actual theremin, the film employs "James Steven's theremin-like flutes" as part of the underscore, which is made up of "cues taken from Britain's Impress Mood Music Library" (Mandell 1996: 2).

3 On "Good Vibrations," Brian Wilson brought in an Electro-Theremin for his studio work. Paul Tanner had invented the Electro-Theremin to mimic the theremin while being easier to play, and Tanner himself produced the famous Electro-Theremin bits that make "Good Vibrations" instantly identifiable (Boilen 2013).

4 I refer here to the discussion between J.L. Austin and Jacques Derrida, in which Austin argues that performance is not performative, but rather quoted "in ways parasitic upon its normal use-ways which fall under the doctrine of the etiolations of language. All this we are excluding from consideration" (Austin 1962: 22). For Derrida, however, there is no origin of the performer to be found, but rather a *différance*, a "spectre of something else" (Derrida 1988, 2005: 151), or as Peggy Phelan writes how: "Any search for an origin is hysterical" (Phelan 2014).

5 For instance, in the United States, the music and dance of the slaves' cakewalk mocked the slaveholders' own music and dance in a way the slaveholders mistook for a primitive and flattering tribute (Sundquist 1993: 278).

References

Austin, J.L. 1962. *How to Do Things with Words*. Cambridge: Harvard University Press.

Barad, K. 2003. Posthumanist Performativity: Toward an Understanding of How Matter Comes to Matter. *Signs: Journal of Women in Culture and Society*, 28 (3), 801–831.

Barad, K. 2007. *Meeting the Universe Halfway: Quantum Physics and the Entanglement of Matter and Meaning*. Durham: Duke University Press.

Boilen, B. 2013. The Real Instrument behind the Sound in 'Good Vibrations'. In *All Songs Considered*. National Public Radio. Accessed July 11, 2025. https://www.npr.org/sections/allsongs/2013/02/07/171385175/no-it-wasn-t-a-theremin-on-good-vibrations-remembering-paul-tanner

Braun, H.-J. 2009. Pulled Out of Thin Air? The Revival of the Theremin. In: K. Bijsterveld and J. van Dijck, eds. *Sound Souvenirs: Audio Technologies, Memory and Cultural Practices*. Amsterdam: Amsterdam University Press, 139–151.

Chude-Sokei, L. 2016. *The Sound of Culture: Diaspora and Black Technopoetics*. Middletown: Wesleyan University Press.

Combes, M. 2013. *Gilbert Simondon and the Philosophy of the Transindividual*. Translated by T. LaMarre. Cambridge: The MIT Press.

Cox, B. and Forshaw, J. 2012. *The Quantum Universe: Everything that Can Happen Does Happen*. London: Penguin Books.

Cull Ó Maoilearca, L. 2017. Equalizing Theatre and Philosophy: Laruelle, Badiou, and Gestures of Authority in the Philosophy of Theatre. *Performance Philosophy*, 3 (3), 730–750.

Derrida, J. 1988. Limited Inc a b c ... *Limited Inc*. Translated by S. Weber. Evanston: Northwestern University Press.

Derrida, J. 2005. *Paper Machine*. Translated by R. Bowlby. Stanford: Stanford University Press.

Dijck, J. 2009. Remembering Songs through Telling Stories: Pop Music as a Resource for Memory. In: K. Bijsterveld and J. van Dijck, eds. *Sound Souvenirs: Audio Technologies, Memory and Cultural Practices*. Amsterdam: Amsterdam University Press, 107–119.

Eidsheim, N.S. 2015. *Sensing Sound: Singing and Listening as Vibrational Practice*. Durham: Duke University Press.
Ettinger, B.L. 1997. The Feminine/Prenatal Weaving in Matrixial Subjectivity-as-Encounter. *Pscyhoanalytic Dialogues*, 7 (3), 367–405.
Ettinger, B.L. 2001. Wit(h)nessing Trauma and the Matrixial Gaze: From Phantasm to Trauma, from Phallic Structure to Matrixial Sphere. *Parallax*, 7 (4), 89–114.
Ettinger, B.L. 2002. Trans-Subjective Transferential Borderspace. In: B. Massumi, ed. *A Shock to Thought: Expression after Deleuze and Guattari*. London: Routledge, 215–239.
Ettinger, B.L. 2004. Weaving a Woman Artist with-in the Matrixial Encounter-Event. *Theory, Culture & Society*, 21 (1), 69–94.
Ettinger, B.L. 2006. Matrixial Trans-Subjectivity. *Theory, Culture & Society*, 23 (2–3), 218–222.
Ettinger, B.L. 2009. Fragilization and Resistance. *Studies in the Maternal*, 1 (2), 1–31.
Glinsky, A. 2005. *Theremin: Ether Music and Espionage*. Urbana: University of Illinois Press.
Guattari, F. 1984. *Molecular Revolution: Psychiatry and Politics*. Translated by R. Sheed. Harmondsworth: Penguin.
Haraway, D. 2004. The Promises of Monsters: A Regenerative Politics for Inapproriate/d Others. In: *The Haraway Reader*. New York: Routledge, 63–121.
Holbraad, M. 2007. The Power of Powder: Multiplicity and Motion in the Divinatory Cosmology of Cuban Ifá (or *mana*, Again). In: A. Henare, M. Holbraad, and S. Wastell, eds. *Thinking through Things: Theorising Artefacts Ethnographically*. London: Routledge, 189–225.
Laruelle, F. and Ó Maoilearca, J. 2015. Artistic Experiments with Philosophy François Laruelle in Conversation with John Ó Maoilearca. In: C. Cox, J. Jaskey, and S. Malik, eds. *Realism Materialism Art*. Berlin: Sternberg Press.
Laruelle, F. 2000. Identity and Event. *Pli*, 9, 174–189.
Laruelle, F. 2010. *The Future Christ: A Lesson in Heresy*. Translated by A.P. Smith. London: Continuum.
Laruelle, F. 2012. *From Decision to Heresy: Experiments in Non-Standard Thought*. R. Mackay, ed. Falmouth: Urbanomic.
Laruelle, F. 2012. *Photo-Fiction, A Non-Standard Aesthetics*. Minneapolis: Univocal.
Laruelle, F. 2013a. *Principles of Non-Philosophy*. Translated by N. Rubczak and A.P. Smith. London: Bloomsbury.
Laruelle, F. 2013b. *Philosophy and Non-Philosophy*. Translated by T. Adkins. Minneapolis: Univocal Publishing.
Luhmann, N. 1993. *Risk: A Sociological Theory*. Translated by R. Barrett. Berlin: Walter de Gruyter.
Mandell, P. 1996. Forty Year Mystery Solved: The Music Behind Plan 9 from Outer Space. *Film Score Monthly*, 1 (69), 1–2.
Massumi, B. 1998. Requiem for Our Prospective Dead: Toward a Participatory Critique of Capitalist Power. In: E. Kauffman and K.J. Heller, eds. *Deleuze and Guattari: New Mappings in Politics, Philosophy, and Culture*. Minneapolis: University Press, 40–64.
Montague, S. 1991. *Rediscovering Leon Theremin. Tempo, New Series*, 177, 18–23. Cambridge University Press.
Mori, M. 2012. The Uncanny Valley. *IEEE Robotics & Automation Magazine*. Translated by K.F. MacDorman and N. Kageki, 2, 98–100
Phelan, P. 2014. Performance Theory Intensive. In seminar at McGill University, March–April, 2014.
Phipps, K. 2002. Tom Waits interview. *AV Club*. May 29. https://www.avclub.com/tom-waits-1798208228, retrieved June 10, 2019.

Rodgers, T. 2010. *Pink Noises: Women and Electronic Music and Sound*. Durham: Duke University Press.
Shaviro, S. 2006. Simondon on individuation. *The Pinocchio Theory*. Accessed March 30, 2023. http://www.shaviro.com/Blog/?p=471
Star, S.L. and Griesemer, J.R. 1989. Institutional Ecology, 'Translations' and Boundary Objects: Amateurs and Professionals in Berkeley's Museum of Vertebrate Zoology, 1907-39. *Social Studies of Science*, 19 (3), 387–420.
Stoll, R.W. 2006. *Touch! Don't Touch! Music for Theremin*. Booklet text for CD, Wergo, WER 6679-2. B. Buchholz and L. Kavina, eds. Mainz: Schott Music & Media.
Strathern, M. 1988. *The Gender of the Gift: Problems with Women and Problems with Society in Melanesia*. Berkeley: The University of California Press.
Sundquist, E.J. 1993. *To Wake the Nations: Race in the Making of American Literature*. Cambridge, MA: The Bellknap Press of Harvard University Press.
Thompson, M. 2017. *Beyond Unwanted Sound: Noise, Affect and Aesthetic Moralism*. New York: Bloomsbury.
Virno, P. 2004. *The Grammar of the Multitude: For an Analysis of Contemporary Forms of Life*. Translated by I. Bertoletti, J. Cascaito and A. Casson. Los Angeles: Semiotext(e).
Young, L.M. 1991 [1962]. The Second Dream of the High-Tension Line Stepdown Transformer. *The Four Dreams of China*. CD. Gramavision: R2 79467.
Zukauskaite, A. 2010. Biopolitics: Antigone's Claim. In: S.E. Wilmer and A. Zukauskaite, eds. *Interrogating Antigone in Postmodern Philosophy and Criticism*. Oxford: Oxford University Press, 67–81.

8 Cow choirs

Singing-with more-than-human herds

Jennie Tiderman-Österberg

We follow a gravel road. As I walk, my feet are gravitationally pulled toward the vibratory earth.[1] Each step produces a crackling sound when shoes meet sand and small pebbles. I find myself in the middle of many rhythms. Sudden increases of wind that take part in the rustling of leaves, the crackling sound of feet and soil, the comings and goings of clouds that hide the sun and its warmth, and the unpredictability of my breath due to the resistances offered by uphill, and downhill, slopes. Muscles get activated, flexed, and firm in tone with that which is encountered. The skin does not cut "me" off from these encounters. It is not a plate armor; it is permeable and leaves the body open to influences (Bennett 2020). Blood vessels, bowels, lungs, tendons, osteoclasts, neurons, and atomic elements are on the move in a body that walks and breathes. This is vital and vibrant matter that partakes in all activities: I am moved by forces that do not answer to my name, to put Rosi Braidotti's (2022) wording to work.

As I look around, I see a myriad of bodies that are more-than-human: creatures within and around us which make my embodied existence dependent on that which often is othered – excluded from, and oppressed by, the self-proclaimed sovereignty of human subjects (Braidotti 2013; Haraway 2016; Jaque et al. 2020). I see short and long legs, large and small cloven hooves. Their steps create different rhythmical figures that do not follow a pre-given pattern. Sonic manifestations of their cloven hooves pressing upon the sand, together with their differently sized bells clinging for each movement, create a tapestry of tempi and timbres that affect and produce each body in this more-than-human herd.

I encounter eyes with horizontal pupils that inquiringly sweep over the environment, and I meet the gazes of black, deep, and wise eyes surrounded by long and thick lashes that bat in a rhythm slower than my own blinking. I can smell them too, my colleagues in this community of cows and sheep. Smells I can only describe as earthy and … brown. The scents of cows and sheep give me a sense of comfort. I am home.

One of the cows hastens to catch up with me. She licks my hand, and I stroke her head and cheek. Her tongue feels rough against the palm of my hand, but it is a friendly gesture. Her coat is soft and warm, and it speaks

DOI: 10.4324/9781003390992-12

to me through temperatures and touch – tactile perceptions that do not only affect surfaces but dig into the depths of the body to recompose it into something new. We melt together, the cow and I, and the boundaries between us are blurred (Haraway 2013). It is a dissolvement of metaphysical distinctions between self and other, human and animal, body and mind (Kassabian 2013). I also hear different voices. A roar resonates with the forest like thunder, and rhythmical bleats come together as a weave of voices sonically present in different frequencies and paces. Myriads of vibrational particles penetrate the whole body. It is more-than-sound that hits me, a sound more-than-audible (Eidsheim 2015).

It is by the grace of this lively sensescape of intertwined, affective flows, that I sing (Figure 8.1). My voice is co-produced by everything that I am not: connections to the gravel road, the sand, the wind, the foliage, the voices of sheep and cows, the body attuned to resistances offered by slopes and mires, the roughness of a tongue sensed through surface-on-surface kinesthetic tactility. Voice is produced by more-than-human relations and materiality. This is what the present chapter will explore further, with a main focus on human-cow relations.

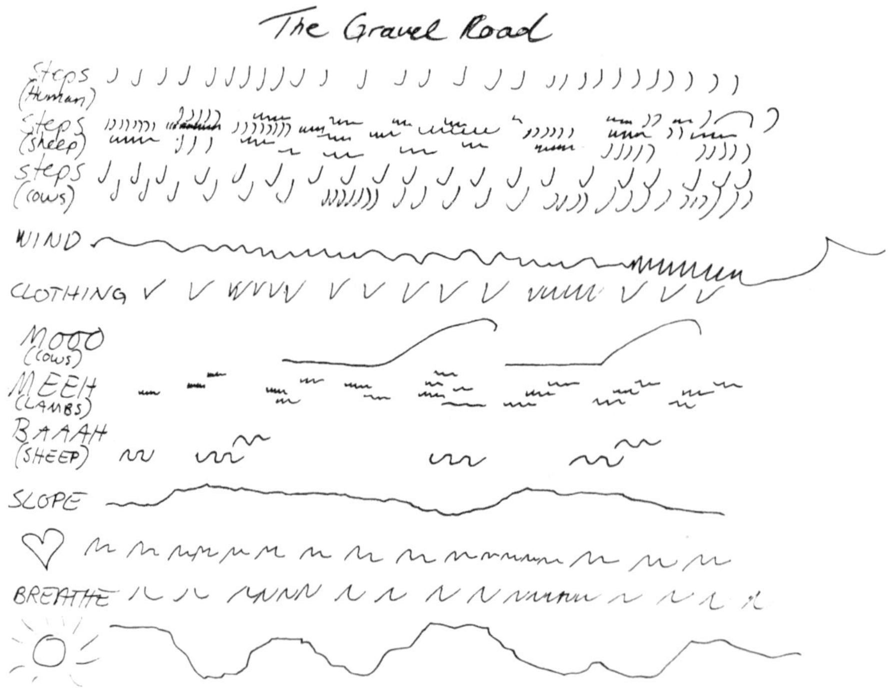

Figure 8.1 Author's fieldnotes. Traveling lines, not a transcript. The tapestry of rhythms of the gravel road.

Entering the field in the middle

Herding music, or herding *musicking* (Small 1998), refers to musical activities[2] that are part of that which constitutes herding cultures such as Nordic *fäbod* or *seter* farming – a transhumance culture where farmers transfer to satellite farms during the summer seasons to make use of the nutritious forest pastures. During the stay at these farms, animals grace freely and unfenced. The herders, who historically were women, developed a whole musically enriched communication system to reach out across large expanses of space. Sounds from cow-, buck, and wooden horns were, and are, used to communicate with cows, goats, and sheep, as well as with other herders and farmers. In addition, a vocal technique commonly named *kulning* emerged with the same communicative functions as the sounds from herding instruments (Tiderman-Österberg 2020; Stomberg and Tiderman-Österberg 2021; Tiderman-Österberg forthcoming). Today, *kulning* has traveled from the farms into popular culture and artistic stage performances (Rosenberg 2019, 2013; Tiderman-Österberg 2019a, 2019c). Even so, hidden from public eyes, *kulning* is still sounding in the forests which embed the fäbod farms (Tiderman-Österberg 2019b). When I started to study and perform *kulning*, the vocal technique and the learning of traditional melodies were the center of my attention. I sought to imitate recordings, and sing, or call, like others. However, it is not the human teachers who have educated me the most about herding musicking. I have learned the most important lessons from cows. They have taught me that it is of no matter how beautifully I sing, nor how perfectly I could represent traditional melodies. In fact, it is not all about me. What matters are respectful and democratic human-animal relations, openness, and understanding. This taught me to be humble and to question the imagined centrality of humans and the stubborn idea that only humans make music and art.

This stirred me in new academic directions toward eco- and zoömusicology, as well as posthumanist and new materialist perspectives on music and voice. Scholars in ecomusicology generally think with ecological perspectives, as well as ethical approaches regarding equality between human and more-than-human entities in music-making practices (Allen and Dawe 2016; Edwards 2016; Seeger 2016; Simonett 2016; Silvers 2020). Ecology, or ecological, are, of course, not neutral, uncontested concepts in academia. They may refer to any kind of infrastructure or interactive system, but here, in this chapter, they refer to nature-cultural entanglements with an emphasis on relationality that stretches beyond human subjects by not drawing a sharp line between nature and culture, or human and animal (Allen 2011; Braidotti 2013). The growing field of zoömusicology, with Steven Feld (2012) as an intensive early source of inspiration, explores musicality among nonhuman animals (Doolittle 2007, 2008; Rothenberg 2001, 2013; Mundy 2018; Taylor 2017, 2020). Studies in both eco- and zoömusicology, or in ecocritical musicology, have shown how music research may produce knowledge regarding the state of the human in relation to all other things that live. Included in this project are, of course, also posthumanist and new materialist musicologists and scholars in artistic research who specifically explores the relationships between nature and culture, and human and nonhuman, to dismantle these dichotomies by emphasizing more-than-human relationality down to the particle level. Of great interest to my research are advances made in voice studies and ethnomusicology that zoom in on voice as an effect of entanglements (assemblages) of more-than-human kinds (Moisala 2017; Moisala et al. 2017; Tiainen 2017a, b, 2018). Forwarded in these studies is the idea that voice, or more specifically, a throat in utterance or in ex/inhalation, arrives because a speaker, a singer or a breather (human or otherwise) always stands in relation to (and is produced by) other heterogeneous elements: vibrating particles, clothing, mucus, bacteria, gravel smells, wind, nonhuman animals landscape, technology, media, art, and more (Neumark 2017; Arlander 2019; Beahrs 2021). By this, a voice, human or otherwise, is always many: simultaneously performative, vibrantly material, tactile, and discursive – some productive elements silent and some sounding all at once (see also Belgrano 2020; Cavarero 2005; Chadwick 2020; Magnat 2019).

By thinking with ecological, posthumanist, and new materialist perspectives, I focus on Scandinavian *vocal* herding musicking, such as long-distance calling (*kulning,* or rather, *to kula,* the verb form), as well as low-pitched songs and humming while milking and herding. My point of departure is that herding musicking is about more-than-human communications, interactions, and affective transmissions. Following from this, my view is that herding musicking is a multispecies activity, not a human solo act. We, the posthuman herders, call *with* more-than-humans. We speak and sing *with* them. In this chapter, I seek to understand this *with: how* multispecies relations propel vocal herding musicking and how affect (connective intensity and energy) works in those relations? This is an exploration of the processes of production; how calling, singing, or humming is continuously organized by more-than-human relationality and materiality.

Now, what is required theoretically and conceptually to explore voice, in particular vocal herding musicking, from a more-than-human perspective? First, it matters what ideas we use to think about other ideas, to quote anthropologist Marilyn Strathern (as cited in Haraway 2016: 12); it matters how we see nature, animals, plants, humans, and of particular interest in this book, matter and affect.

Animals, affect, sympathy

During the past decades, a plethora of literature has been published that takes interest in posthumanism, environmental ecologies, and new materialist, more-than-human perspectives on music and politics (see, e.g., Bignall et al. 2019; Åsberg et al. 2011, 2018a, b; Tiainen, 2018; Gruen et al, 2018; Moisala et al, 2017; Thompson and Biddle, 2013; Braidotti, 2013). Scholars in these fields have emphasized that we need to be confronted with the fact that humans always stand in relation to other species and forces that affect how actions are produced, and lives are lived. This means that we also need to rethink music as an activity and capacity that is not uniquely human. Researchers in zoömusicology, and in human-animal studies, have pointed out that nonhuman animals communicate, speak, feel, use tools, and make music (Dolittle 2007; Dagg 2011; Krause 2012; Mundy 2018), and they, too, are vulnerable and mortal, as well as individually different beings (Wolfe 2009; Corman 2020). Most importantly, animals are not seen as objects or "the others" here. Rather, they are vagabond actants (Bennett 2010) in our lives. Companion species are, as Donna Haraway (2008) writes, "fleshy, significant, semiotic-material beings" (p. 165). These vagabond actants (think not only of pets here but also of the smallest of insects and bacteria, see e.g. Pedersen 2011), or full partners, are always everywhere (Haraway 2003, 2008: 301), co-producing every aspect of our lives, from activities to gazes and places (see, e.g., Birke and Holmberg 2018; Bull and Holmberg 2018). For example, when I see an animal and conclude that it is a cow that I see, my view is already co-produced by her. Even anthropomorphism is not completely my doing. Some aspects of us, may it be just olfactory elements or bacteria, will always melt together with those of other bodies.

My point of departure, then, is that animals act alongside, or even upon, us because bodies of various kinds sympathize with each other in constant communication (Plec 2013; Menely 2015; Schötz 2019).

Even though this chapter focuses on human-animal relations that give rise to singing, it is worth emphasizing that animals are not the only vagabond actants that affect our lives and actions. There are also elements that do not belong to the animal kingdom, namely the lively flora that enfolds, and connects to, bodies in action (Pitt 2017; Arlander 2019; Bennett 2020). We might also encounter technological elements that influence our actions, but also the actions of other organic bodies (Kember and Zylinska 2012). As we all know, we might also be forced to change everything in our lives due to a virus.

When approaching herding musicking as singing-with, a communicative activity born out of human-animal relations shaped also by other vagabond actants, language-based explorations will not suffice. The cows and I do not share the complex meanings of spoken and written "human" words, sentences, and concepts. What we do share are capacities to affect and be affected (Nyman and Schuurman 2016), or rather the capacities to affect-*while*-being affected. Bodies have capacities to affectively *attune* to other bodies, human or more-than-human, through affective "circuits of sympathy", to use political theorist Jane Bennett's (2020: 27) wording, where sympathy is not sentimental pity felt by an individual, but an "atmospheric force" (Bennett 2020) that is more about vague tones that are produced, encountered and felt viscerally, than about perceptions named by "those who undergo them" (Bennett 2020: 24). This perspective requires two things. First, that matter

Figure 8.2 Humanimal embrace. Melting into goats.
Source: Sketch by author.

is active and affective, and second, that affect is about more-than-human emotion or the feelings of an individual. Rather, affect is in this chapter approached as energetic flows between bodies – flows that also run *through* bodies, transforming (producing) them as they do (see also Manning 2009).

My approach to communication is about more than an exchange of information that is somehow transmitted through a medium, then unpacked and made meaningful. It is, as Emily Plec (2013) stresses, the capacity to *feel* with another that enables communication and, ultimately, perception (see also Menely 2015). All animals, including humans, communicate and speak through scents, vibrations, body positions, and movements, or even through vague pressures and energetic nudges as part of ordinary, everyday life (Stewart 2007; Manning 2016). Therefore, affects can be approached semiotically where intensities become vibratory and fleshy non-institutionalized micro-signs that cross borders between species and challenge the idea that bodies are mute lumps of flesh that need to be made intelligible to be communicative (Coole and Frost 2010; Stark 2017).

These affective sympathetic circuits are also closely explored by Cynthia Willett (2014) as she dives into the world of *eros* (p. 9), where erotica is not the sexual or romantic desires of the self but about responsive ethics of compassion risen from "bodily and sensory immersion in a partly shared world" (p. 11). The focus is not on what sets humans and other animals apart but on what makes us the same.

However, this is a utopian world. As Donna Haraway (2008) warns us, the interest in affect as visceral and non-subjective intensive energies that connect and produce bodies involves certain risks. It might, for example, strengthen the idea that nonhuman animals are not capable of a sense of selfhood and individual sensations and feelings. Therefore, we might miss out on close, personal, and intimate (and thereby affective) relationships that we have with other animals. As Braidotti's (2017) critical posthumanism teaches us, when we decentralize the liberal human subject, we also need to be careful not to lose our sight when it comes to the workings of power and hierarchies that are possible threats to bodies' capacities to act. Certain expressions of domination matter for the workings of affects. This is, of course, of great importance when working with human-animal communication. To not get stuck in a "dreamlike life", as Willett (2014: 11) writes, she puts Daniel Stern's ideas of categorical and vitality affects to work (see also Gibbs 2010; Volgsten 2019). Vitality affects are here approached as affective contours that are prerequisites for conversations between bodies that do not share verbal language, such as an infant and caregiver, a human and an animal, or a cat or a cow. Here, vocal timbres, gestures, and movements function body-semiotically even though they do not require conceptual thought to be communicative. Categorical affects, however, have names and could be ordered as good or bad. These affects are also about those who undergo them, their previous relationships, and their situated experiences. In this way, different affects *cross-fertilize* (my wording) each other. There are dissonances and harmonies at work. Disturbances and ruptures, as well as visceral and subjective surfaces, intensifications, and recessions, all at once. I will develop these ideas further when I explore the empirical material, but first, few words on methodology and methods.

How to travel unfolding paths

Our onto-epistemological (knowledge as a relation-in-production. See Barad 2007), conceptual, and ethical approaches matter for the selections and methods we, as researchers, make regarding what stories we should tell, what voices we should listen to, and what signs, gestures, and vibrations we should interpret as meaningful. Put differently, these ideas affect how we perform research, and how we study affective voices, as more-than-human music scholars who take an interest in new materialism.

Affect and sympathy are concepts that will be put to work in ethnographic, or rather cartographic, explorations in this chapter. The cartography, that is, a mapping of affective flows, is produced poetically and affectively, propelled by the intensities of different encounters (Stewart 2007; Bennett 2020). I seek to follow affective energies and currents, but also strong ruptures and minor transformations, that organize and produce singing and calling, or obstruct these activities. Following Chadwick (2020, see also Mazzei 2009), the mapping of the production of voice as a movement produced by more-than-human relationality and sympathy needs to be explored through a "multisensory engagement" (p. 5) with "fleshy voices" (p. 5) and "embodied rhythms" (p. 5). This means that I do not seek to present exactly what is sung (through, for example, Western staff notation), but rather *how* something is sung and trace the relational processes of production (affect and sympathy) behind it. To incorporate bodies as more-than-text, I use drawings as part of the cartography. I name this addition *sketchography*. Activated by Stephanie Jones and James F. Woglom's graphical scholarship (2015. See also Lünenborg and Maier 2019; Yunkaporta 2020), it is my conviction that drawings might add a level of fleshiness and bodily aliveness to carto- and ethnographies. To put it simply, the affective shapes, postures, and movements of bodies are more easily retained in drawings than in words. Included in sketchography is a type of transcription of voices that seek to line affective contours: the timbres, dynamics, intervals, and rhythms of vocalizations, rests, and breaths.

The cartography explores empirical material (fieldnotes and sounding video recordings) produced during three different periods of my fieldwork in 2020. When the material was produced, I always made sure to not only make video recordings with my human colleagues but also with many "eco-others" (Braidotti 2011, 2013). I will take the opportunity here to fathom my own experiences of singing-with sheep and cows. It is a performative, practice-led exploration (Haseman 2006), where the sheep, cows, and I together produce knowledge about herding musicking as singing-with. The approach that I am offering here is by this also connected to more-than-human participatory research (see also Bastian et al. 2017; Silvers 2020) with an ethical sensitivity to acts of domination and power.

Let us leap into the field as it unfolds.

Exploration I: The gravel road

The old ram takes it slow, and his voice is deep and quite hoarse, although fiercely loud. The lambs are playful, and they call out in light, fast rhythmical figures. One cow is at the front of the herd, and she is calling in long sentences, starting with a low frequency that subsequently rises to a higher one as the voice grows thinner.

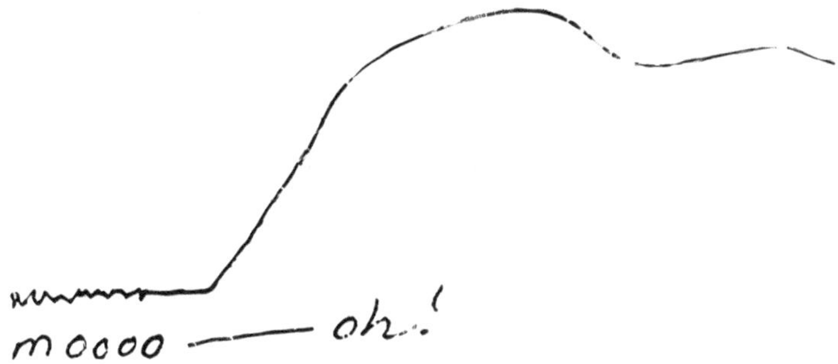

Figure 8.3 Cow's roar. From a thundering murmur in the belly to a loud, but thinner, call.

Source: Author's fieldnotes.

As the frequency rises, her head is lifted, and she gazes straight ahead. The open landscape invites her long call and her far-reaching glaring. It is like the contour of the long winding road haunts the phrasing of her roar and the movements of her neck, head, and gaze. Her nostrils are flared. Are there dangers here? She has also been here before. It shows in the depths of her eyes. Memories are activated at the same time as new encounters stream through her like flashes. Like there is a resonance between memories and new encounters.

The gravel road journey is a togetherness of atmospheric affect and sympathy between a myriad of elements: proprioceptive feedback, limbs, motor commands, neurological circuits, slopes, sand, stones, bones, rubber shoe soles, and gravity. These elements all take part in the walking, organizing steps, and this activity's sonic trails. I am walking because the connections between these elements are alive. Stock and stone, inclination and air, produce this body that walks.

During our journey, walking is accompanied by singing and calling. Short phrases it must be since I am breathing heavily when encountering an uphill slope. Here, voice takes part in, and attunes to, the dynamic soundscape of sonically manifested life. As bodies recompose in dialogue with the surroundings, voice is affected and reflected by different surfaces. This means that vocal frequencies and rhythms emerge with the soundscapes, or sensescapes, of the vibrating particles of inaudible sound and body parts already on the move, plugged into, and in conversation with the fleshiness of voice. Voice is also transforming due to more-than-human bodies on the move, and the dynamism of their affective intensity.

Four cows are lagging. I do not see them, but I can hear and feel them behind me. Their breathing, their steps, their smell, and their energy are presences felt as a pressure on my back and on the senses. I turn around, calling: "come on now". The temperament of the call is not demanding but a search for contact, a desire to plug into the laggards in a more intense way; a connection that might glow stronger than the loudness of wind and the resistance it offers. The last two of them ignore the call, and voice transforms almost-before this ignorance is manifested. It is as if the intensity of ignorance produces a new, sturdier, voice. Intensity, as aspects of other bodies, intensifies: “Hello!? Come on now! Ko-ho-di-di-ja! Come on now!”

Interestingly, walking stops when the contact is weakened. Yet, the voice flows. Rupture happens in one mode (walking), while flow continues in another (singing and calling). My voice grows thinner when I see one of them. Her head is lowered, and she walks slowly. Suddenly, I feel as if all the energy has left my body. This affects voice, which grows thinner, and I say in a friendly voice “Come here now”. The cow’s lowered head and her slow pace organized voice through mimesis. Aspects of one body flow in and through another to alter it. The effect is the transformation of voice.

As the last shirkers, let us name them “Felice” and “Lilly”, draw closer, yet not nearly beside me, walking starts again. It is almost as if the decreased distance between us creates a new incitement for walking. Their almost-movement moves me to walk; it quickens and intensifies the connections that allow me to walk and affect how I walk. After a few steps, something gets me to slow down. I have felt them close to me, but our connective intensities have faded. I turn around only to see that they are left behind again. Calling starts *as* I turn, not after turning is completed: “Come here now!”. Walking slowly turns into rest, and I stop. “Come on here now!” I watch them get closer, and as they pass me, I stroke their coats, pet their backs, speaking to them softly “Well done, *kuan*, (swe. vernacular speech for “cows”), there you go”. While this act is one of shameful human domination, there will always be resistances. We walk together for a while. One of the cows says something. A low and quite muffled “mmmm” that does not end before there is an answer: “Yes”, I say in a long and elastic way. The rubbery movement of her voice returns in mine because we stand in sympathetic communication.

It is like the embrace in art theorist Erin Manning’s (2009) dancing, where proprioceptive feedback and affect as virtual potential propel movements, but there are voices instead of arms in this encirclement. The pace and rhythm

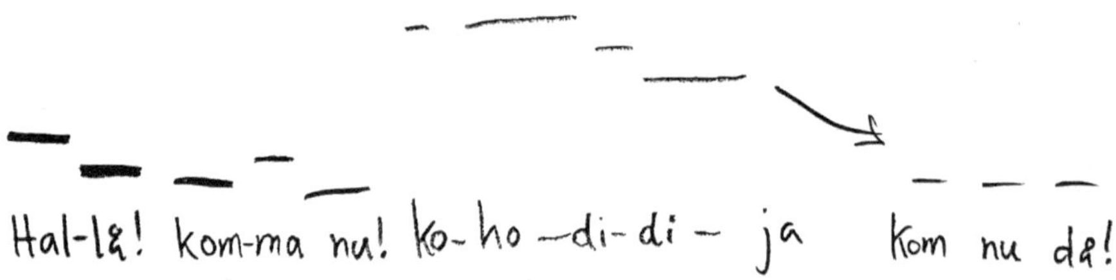

Figure 8.4 A clear annoyance in the voice at first, a full-bodied “hello”. The voice grows thinner, but higher in the melodic call in the middle that ends on a falling note. The section ends with a soft “come here now”. Lyrics in Swedish.

Source: Author’s fieldnotes.

Figure 8.5 Cow with lowered head.
Source: Author's fieldnotes.

of my steps attuned to those of the cows and sheep is the beating of a more-than-human kind of music. A few steps more, and I start so sing a word-less tune. As I watch the video afterward, I notice that the cow bells create an open chord. Fundamental and fifth. Singing begins at the fifth, to drop in an ornamented, improvised melody with the same beat as the pace by which we walk: melodies attuned to cowbells, phrases attuned to walking, voice attuned to breathing, breathing attuned to resistances. Is this not music?

There on the gravel road, herding calling is a voiced connection. Almost-movements, and not yet actualized utterances, come with intensities and potential further actions. Weakened affective bonds and diminishing intensities between bodies are movements that hinder actual movement: walking stops when these bonds grow weaker. The imperceptible but vaguely *felt* minor gestures (Manning 2016) compose differences in calling from within. The minor tendencies, the becoming, produce differences that surprise. A voice in transformational singing has, as Manning (2016: 1) writes, a *wildness:* a bodily attunement of vibrational flesh. Softness is composed when nearness is felt, frustration is voiced when bonds are exposed to disruption. Singing is produced by *eros,* the *with* in *singing-with.*

Exploration II: Tradition becoming vagabond

Milking by hand entails sitting on a small stool made *with* old wood. My forehead rests on Flora's belly. Bodies meet when surfaces attach to one another, and temperatures communicate through the thin linen of my headscarf.

Figure 8.6 Author's drawing. Walking-with cows.

Since I am quite tall, my body is folded into a quite uncomfortable position to reach the udder. The surface-on-surface tactility between my hand and the udder is one plugged-in-ness of the activity. Our bodies together. They need to be together for milking to be possible.

After a while, I start to sing a familiar song, "The Longing of a Fäbod Woman", and it is as if the singing increases the flow of milk. The teat gets fuller, a tension in the skin which propels me to continue to sing. This singing was not planned, nor was the tune chosen beforehand. Yet, it is a song that is traditionally connected to this farming culture, the song of a fäbod woman. It is not strange that this song flows out of my mouth: it traditionally belongs here. However, it is more interesting to explore why I started to sing in that moment, and how the singing transformed the milking activity and vice versa. The rhythm of milking and the flow of milk *move me* into a temporal state of singing. I sing, attuned to the movements of milking. I sing almost in the same rhythm as my hands move, pulling and squeezing, yet slightly off due to a bodily attunement to Flora's breathing and chewing, which are

Figure 8.7 Author's drawing. Milking by hand, always-with Flora.

not completely in phase with the rhythms of my hands. Different rhythms in different body parts, differently attuned dependent on the flow and what was already vibrational in those body parts. I also need to breathe more often than usual, due to my bent body, where there is not much room for the expansion of lungs and diaphragm. I sometimes need to breathe before the ending note of a phrase. As I inhale, the pace of milking decelerates. Breathing and pulling work together, movement and slowness produce voice. Body position and disposition matter, and prehension, the ability to know how to reach and grasp (Grafton 2010), matters. My ability to adjust the technique in tone with the softness of Flora's skin and the density of the teat matters for how singing is produced. When the stream of milk gets thinner and the teat gets emptier, singing stops. It was, then, a kind of singing deeply entangled with milking. Milking was its engine, and without that engine, singing is no more. I change to another teat.

I am humming this herding song without a steady ground pulse, breathing as a part of the rhythmical co-engagement of milking and singing. The steady pulse of the traditional melody, its time signature, is transformed into a singing that works as a line, divided by organic phrases. The division of phrases comes from other movements of the activity. The connection of movements produces phrases and intermissions in singing. Bodies attune and transmit affects between them. Affects saturated with different temperatures, tensions, olfactory, and haptic intensities. We do not attune to a beat which is given to us by a pre-existing framework. It is not entrainment. It is affective attunement that works independently of such patterns. Here we have what Manning (2016) has framed as the minor working the major from within. Minor gestures, the difference in singing due to affective attunement, transform the major singing traditions as framed by recordings and authoritative discourses. Through the minor gestures, meaning that which is communicated before something is uttered, tradition gets transformed. Affected bodies, and lively matter, matter in how a musical tradition is materialized differently. Melody, pitch, and rhythm are all altered due to that with which I am engaged. *How* I sing this song in this moment in time has to do with more-than-human culture and tradition. This is tradition becoming vagabond: emergent and in transformation as herders sing *with* cows.

Exploration III: Dissonance and harmony

I can hear her breathe and chew. For some reason, it makes me smile: her sounds are activating joy. I take a seat in the empty booth across the aisle from Nelly's. Her head turns, and she is looking at me. Her almost black eyes inspect me for a long while. I feel dissonance. Is she nervous? Her breathing is intense, especially the exhalations. They are sounding warmth. I start to talk to her with a soft low-pitched voice, telling her how beautiful she is. I start to hum an improvised melody in a low register. I must be low since my vocal cords are worn out, and I feel hoarse. The swollen fleshiness of my voice clearly affects how I sing at that moment: low humming, not a pure and distinct vocalization. Voice is here produced by the state of mucus membranes and fleshy surfaces that are more related to that which Braidotti (2010) names *zoē* – the animalistic part of life that ignores any signification, a life that survives the death of any subject (Braidotti 2010: 206–207).

As humming vibrates through my throat, Nelly looks at me, now holding her breath. She exhales, turning around to face the wall instead of me. I do not know if this is ignorance or fright. I call her name gently, and she looks at me again. I ask her if I can come closer, and I calmly start to move toward her. I walk into the booth next to hers, where they keep the hay. Singing starts again, tranquil and low-pitched, and she looks at me again. I reach out to her with my hand, and she allows me to stroke her neck and cheek. Then she starts to lick my hand. I know that social licking is a way to communicate acceptance, safety, and friendships (see, e.g., Jensen 2017) among cows, and I take this as a sign of acceptance, and I slowly walk into her booth, singing and stroking her belly. I reach out and take a small stool, and I sit down beside her. I am stroking her belly, tilting my head against her, with one hand on her back. Singing in a very low-pitched voice: a long “mmm”.

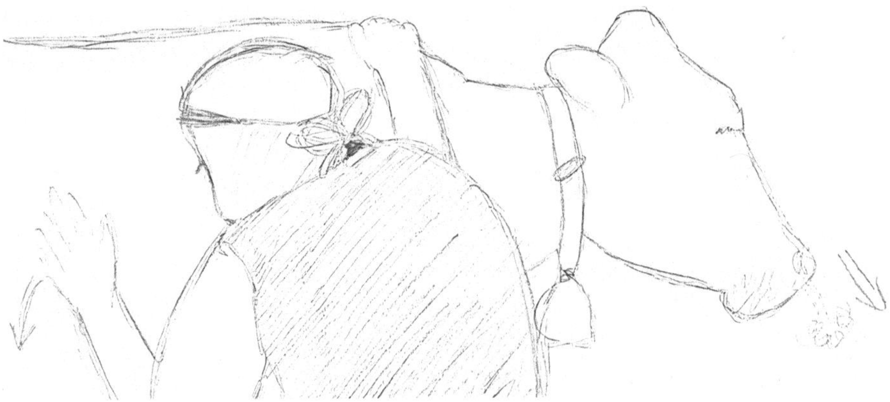

Figure 8.8 Author's drawing. Stroking propels singing.

It is an improvised singing, yet within the Nordic herding tune mode– a general character trait of traditional herding musicking and early Swedish folk music (Ahlbäck 2019, 2020), with the third in-between major and minor, the fourth dynamically transformed in pitch. I am humming without a steady pulse. Continuously stroking Nelly in a movement that is attuned to our breathing and singing, which, in turn, is attuned to the stroking. Our singing and stroking make her almost fall asleep, breathing more slowly, a more protracted exhalation, continuously chewing. After a while, our inhalations and exhalations are almost synchronized, organic, not in perfect harmony. As I sing, she exhales.

I do not force Nelly to exhale with every phrase, inhale as I inhale. Nor do I hold my breath in wait for her exhalation, since singing starts a millisecond before she exhales. As this activity unfolds, we become almost fully synchronized, communicating the almost-exhalation, the almost-singing, before their actualization (as almost-movements in dance are communicated before being actualized, see Manning 2009). I believe that this is an example of affective attunement, bodies communicating through temperatures, postures, gestures, the minor-movements of almost-movements, and the transmission of air.

Now, something happens to Nelly's nervousness during this time of musically enriched togetherness. She relaxes. The change in rhythm is what makes me look at her face, only to discover that her eyes are closed. What is communicated here, through an affective bond, is the reduction of stress, tenseness becoming relaxation. This is communicated before I see that she is calmer. I can feel and hear the transformation of mood. This singing-with-cow is a herder-cow melting together (Haraway 2013). This melting together interlaces with traditional modes of herding musicking. The scale might be a pre-established path, but it does not determine how herding musicking comes to be in this moment. The scale is altered as utterances and phrases are co-produced by a herder-cow relation, and everything else that takes part in this relation (the other cows, the smell of manure and hay, the clinging of cowbells, clothing, ideology, and gendered traditions rise from the fact that fäbod farming and kulning were women's work). This is yet another example of a very wide range of elements and their relations, which give rise to the practice of herding calling and musicking.

•••

We have seen examples of sympathy or eros. Bodies in correspondence, explored by Anna Gibbs (2010) as mimetic communication,[3] or what I would call more-than mimicries. As bodies are affected, they sometimes seem to "imitate" each other's movements and actions; it is a kind of both voluntary and involuntary corporeal attunement, like when we yawn when another yawns or scream out of fear because another screams (see also Thompson 2013). This happens in many dimensions, from visceral and "primitive" ones to individuals engaging in an emotional convergence. Mimetic communication, incorporating mimesis, can be seen as a form of asymmetric, not deterministic, "synchrony" (Gibbs 2010: 187) where an affective sharing is the foundation in all processes of communication approached as transmissions of aspects of bodies to another (Gibbs 2010: 193). Voice as movement can briefly resemble (not copy) another voice due to the workings of affect. Voice is both affectively produced and sonically manifests affect-as-itself (Fast 2018. See also Lochhead et al. 2021).

Mimesis is a multimodal correspondence where it abstracts some aspects of another body. It works cross-modally, where vocalizations and body movements attune to each other affectively because each movement (voice included) has an affective contour that is translated into a different mode: the rhythms of walking might be translated into the rhythms of singing, for example, and each new utterance, each new movement, partakes in the production of what comes next.

These contours affect us very differently, dependent on the state of the affected body and previous relations. This is a sort of affective resonance where some affects are amplified and some muffled depending on body historicity (see also Mühlhoff 2019; Volgsten 2019 as a comparison).

Who I become here, is what *we* make of me. I came to the field of my research with an idea: I am a human, I should control animals. However, this anthropocentric dominatrix was challenged. My selfhood is transformed and made anew. Humbled and touched, I now feel ashamed of my humanness as my throne was crumbled beneath cloven hoofs. Yet, I take my seat on the milking stool to take what is not rightfully mine.

Vocation: Call as if the world matters

What can we make of the explorations above? Do animals matter in singing? This chapter has shown that they do. My explorations have brought forth that herding music is an *emergent* activity co-produced by more-than-humans.

They have also shown that voice is fleshy and physical, constituted by various materialities at the same time as it is discursively produced. This fleshy voice is deeply embedded.

Vocal sounds are never disconnected from the activities that generated them (Eidsheim 2015), and to intonate is not just about listening. It is an embodied activity where vibrating bodies take part in how sound, rhythm, and frequencies move us to attune to their already-aliveness. In a posthuman sense, music and voice are unthinkable without more-than-human, always embodied relations (see also Mundy 2018). It is as music philosopher David Rothenberg (2001) argues: "we cannot make any kind of music without sensing its resonance in an environment" (p. 4), and we cannot make music without nature since all who make music (animals included!) are possible because of that nature (Rothenberg 2001). Nature takes part in voice as a musical instrument (see also Stomberg and Tiderman-Österberg 2021).

I believe it is fruitful to take these perspectives under consideration when approaching more-than-human musicking as a multispecies language where human discursive elements, such as musical traditions, are joined by the animals in and beyond us. From this, we may approach herding musicking as a *vocation:* "calling, calling with, called by, calling as if the world mattered, calling out, going too far, going visiting" (Haraway 2016: 126) – a visit to unfamiliar and potentially unknown futures.

Eulogies becoming embryonic futures

What is ultimately the meaning of engaging in posthumanist projects in voice and music studies in all their philosophical, sometimes speculative, and abstract complexities? Well, as Braidotti (2010) puts it: "For no reason at all. Reason has nothing to do with this. Let's just do it for the hell of it and for love of the world" (p. 217). I want to take up on the last part of this quote, namely, love of the world.

My stance is that ideas that maintain an ontological difference between nature and culture, humans and animals, have played a crucial role in nourishing and reinforcing processes that have led to the oppression and mass extinction of species and the climate crisis. How we see animals and other others matters for what it means to be human. It matters what counts as meaningful and productive. It matters what lives we include in moral and ethical spheres.

Therefore, we should not think, "what is the point", but "this is the point where I depart into something new". This is the point where I believe a little bit more in new materialist voice studies and posthumanist projects, and a little more in critical ecomusicology as fields of potential for sustainable futures, built upon sustainable relations beyond the human. Vocal more-than-human musicking can be one entry point to these projects, among many. By rethinking music as something not uniquely human (who benefits from that kind of distinction anyway?), by allowing all eco-others into moral spheres, by seeing nature as relation instead of resource, by acknowledging our deep and rooted entanglement with that which we have not allowed into the human realm, we might begin to build the micropolitics needed for fundamental transformation toward sustainable futures. Therefore, in further research on these matters, continue to ask what produces and populates voice and breath, and how voice is a material-discursive relation of more-than-human kind. Continue to acknowledge the importance of animals and bacteria, flesh, mucus, and the lively and connected bodies that sympathetically sing. Ask what animals see, hear, taste, smell, perceive, or what they are affecting and affected by. These are the most important eruditions taught to me by cows and gravel roads.

Acknowledgments

The author would like to express a deep gratitude toward the financiers of this research project: Dalarnas museum, Royal Swedish Academy of Music, The Royal Gustavus Adolphus Academy, and Irisstipendiet. She would also like to thank the editors of this volume for providing helpful comments and suggestions for improvements. Finally, she would like to direct appreciation to all the fäbod cows who taught her more about herding music than any human ever could.

Notes

1 This is a paraphrase of Walt Whitman's writings, discussed by Jane Bennett (2020: 21–22), where he writes that the press of feet upon the earth "springs a thousand affections" (Bennett 2020: 21). Bennett's elaboration plays with the thought that each footstep is also a foot*fall,* since feet are also pulled toward the ground by gravity.

2 *Musicking* refers to music as an activity instead of a noun and a knowable object. Musicking is produced by more than composers, conductors, singers, and musicians. In Christopher Small's (1998) example, the audience, composers, etc., also take part in its production. Small does not, however, mention animals or take any expanded interest in nonhumans and matter as an active force. Musicking can also be conceptualized as *events* continuously transformed and constituted by assemblages of various kinds. This means that music, as an activity, is produced by more-than-human elements such as bacteria, vermin, and living materials, but also territorial assemblages such as discourses and authoritative norms (Moisala et al., 2017).

3 It is worth emphasizing that the starting point for Gibbs (2010) is that there is never only one affect at work, but myriads of contagious affects that flows between and through bodies. Inspired by *both* Brian Massumi and David Stern, Gibbs (2010) seeks to lay out an approach to affect relationally, i.e., how it can be understood through the dynamics of the material-discursive, how affective bodies are re-written by "culture", and how "culture" is re-written by affective bodies at the same time. Bodies transform in relation to other bodies. It is a material-discursively constituted process where affective vibrant matter sometimes is at war with discourses surrounding bodies, like when bodily mucus and flows disrupt and challenge the notion on the body as a neatly pulled-together object resting silently beneath the power of logos. Lively matter thus takes part in how discourses are materialized (see Barad 2007; Grosz 1994), and how culture is transformed. Affect takes place between different states of this transformational process, as well as between bodies at a particle level.

References

Ahlbäck, S. (2019). "Den nordiska vallåtens tonspråk." In *Fäbodlandskap och vallmusik*, pp. 114–121. Falun: Dalarnas fornminnes- och hembygdsförbund.

Ahlbäck, S. (2020). *Tonspråket i äldre svensk folkmusik*. 2nd ed. Stockholm: Udda Toner.

Allen, A. S. (2011). "Ecocriticism and Musicology." In *Journal of the American Musicological Society*, 64(2), pp. 391–394.

Allen, A. S. and Dawe, K. (eds.) (2016). *Current Directions in Ecomusicology: Music, Culture, Nature*. New York, NY: Routledge.

Arlander, A. (2019). "Resting with Pines in Nida–attempts at Performing with Plants." In *Performance Philosophy*, 4(2), pp. 452–475.

Åsberg, C. and Braidotti, R. (eds.) (2018a). *A Feminist Companion to the Posthumanities*. Cham, Switzerland: Springer.

Åsberg, C. and Braidotti, R. (2018b). "Feminist Posthumanities: An Introduction." In Åsberg, C. and Braidotti, R. (eds.) *A Feminist Companion to the Posthumanities*, pp. 1–22. Cham: Springer.

Åsberg, C., Koobak, R. and Johnson, E. (2011). "Beyond the Humanist Imagination." In *NORA - Nordic Journal of Feminist and Gender Research*, 19(4), pp. 218–230. https://doi.org/10.1080/08038740.2011.625042

Barad, K. (2007). *Meeting the Universe Halfway: Quantum Physics and the Entanglement of Matter and Meaning*. Durham: Duke University Press.

Bastian, M., Jones, O., Moore, N. and Roe, E. (2017). "Introduction. More-than-Human Participatory Research." In Bastian, M., Jones, O., Moore, N. and Roe, E. (eds.) *Participatory Research in More-than-Human Worlds*, pp. 1–15. London and New York: Routledge (e-book).

Beahrs, R. O. (2021). "Gifts of the Sygytchy-Sons: Tethering Melodies to Land, Kin, and Life Energy at the Khöömei Ovaa, Tyva Republic." In *Asian Music*, 52(2), pp. 72–107. https://doi.org/10.1353/amu.2021.0015

Belgrano, E. L. (2020). "Mapping the Burden of Vocality. French Seventeenth-Century Vocal Lamentations, Japanese Meditation and Somatic Intra-Action." In Kapadocha, C. and Belgrano, E. L. (eds.) *Somatic Voices in Performance Research and Beyond*. Oxon and New York: Routledge.

Bennett, J. (2010). *Vibrant Matter. A Political Ecology of Things*. Durham & London: Duke University Press.

Bennett, J. (2020). *Influx and Efflux: Writing Up with Walt Whitman*. Durham & London: Duke University Press.

Bignall, S. and Braidotti, R. (eds.) (2019). *Posthuman Ecologies: Complexity and Process after Deleuze*. Lanham: Rowman & Littlefield International.

Birke, L. and Holmberg, T. (2018). "Intersections. The Animal Question Meets Feminist Theory." In Åsberg, C. and Braidotti, R. (eds.) *A Feminist Companion to the Posthumanities*, pp. 117–128. Cham: Springer.

Braidotti, R. (2010). "The Politics of 'life itself' and New Ways of Dying." In Coole, D. and Frost, S. (eds.) *New Materialisms. Ontology, Agency and Politics*, pp. 201–220. Durham & London: Duke University Press.

Braidotti, R. (2011). *Nomadic Theory: The Portable Rosi Braidotti*. New York: Columbia University Press.

Braidotti, R. (2013). *The Posthuman*. Cambridge: Polity.

Braidotti, R. (2017). "Posthuman Critical Theory." In *Journal of Posthuman Studies*, 1(1), pp. 9–25. https://doi.org/10.5325/jpoststud.1.1.0009

Braidotti, R. (2022). "We are rooted, but we flow". Afscheidsrede Prof dr Rosi Braidotti (Farewell lecture). https://vimeo.com/720589834. Link 2022-08-23.

Bull, J. and Holmberg, T. (2018). "Introducing Animals, Places, and Lively Cartographies." In Bull, J., Holmberg, T. and Åsberg, C. (eds.) *Animal Places: Lively Cartographies of Human-Animal Relations (Multispecies Encounters)*. London and New York: Routledge.

Cavarero, A. (2005). *For More than One Voice: Toward a Philosophy of Vocal Expression*. Stanford, CA: Stanford University Press.
Chadwick, R. (2020). "Methodologies of Voice: Towards Posthuman Voice Analytics." In *Methods in Psychology*, 2. https://doi.org/10.1016/j.metip.2020.100021
Coole, D. and Frost, S. (2010). "Introducing the New Materialisms." In Coole, D. and Frost, S. (eds.) *New Materialisms. Ontology, Agency and Politics*, pp. 1–46. Durham & London: Duke University Press.
Corman, L. (2020). "He(a)rd. Animal Cultures and Anti-Colonial Politics." In Struthers Montford, K. and Taylor, K. (eds.) *Colonialism and Animality: Anti-Colonial Perspectives in Critical Animal Studies*, pp. 159–180. London & New York: Routledge.
Dagg, A. I. (2011). *Animal Friendships*. Cambridge: Cambridge University Press.
Doolittle, E. L. (2008). "Crickets in the Concert Hall: A History of Animals in Western Music." In *TRANS. Revista Transcultural de Música*, 12.
Doolittle, E. L. (2007). *Other species' Counterpoint. An Investigation of the Relationship between Human Music and Animal Songs*. Diss. Princeton University.
Edwards, J. R. (2016). "Critical Theory in Ecomusicology." In Allen, A. S. and Dawe, K. (eds.) *Current Directions in Ecomusicology: Music, Culture, Nature*, pp. 151–164. New York, NY: Routledge.
Eidsheim (2015). *Sensing Sound: Singing & Listening as Vibrational Practice*. Durham: Duke University Press.
Fast, H. (2018). "Vocal Nest: Non-Verbal Atmospheres that Matter." In *JAR: Journal for Artistic Research* (16). https://doi.org/10.22501/jar.387047
Feld, S. (2012). *Sound and Sentiment: Birds, Weeping, Poetics, and Song in Kaluli Expression*, 3rd ed., Philadelphia: University of Pennsylvania Press.
Gibbs, A. (2010). "After Affect. Sympathy, Synchrony, and Mimetic Communication." In Gregg, M. and Seigworth, G. J. (eds.) *The Affect Theory Reader*, pp. 186–205. Durham [N.C.]: Duke University Press.
Grafton, S. T. (2010). "The Cognitive Neuroscience of Prehension: Recent Developments." In *Experimental Brain Research*, 204(4), pp. 475–491. https://doi.org/10.1007/s00221-010-2315-2
Grosz, E. (1994). *Volatile Bodies. Toward a Corporeal Feminism*. Bloomington, US: Indiana University Press.
Gruen, L. (ed.) (2018). *Critical Terms for Animal Studies*. Chicago: The University of Chicago Press.
Haraway, D. J. (2003). *The Companion Species Manifesto: Dogs, People and Significant Otherness*. Chicago: Prickly Paradigm.
Haraway, D. J. (2008). *When Species Meet*. Minneapolis: University of Minnesota Press.
Haraway, D. J. (2013) *Donna Haraway on the 'humanimal'*. https://www.youtube.com/watch?v=BUA_hRJU8J4. Link 2022-08-23.
Haraway, D. J. (2016). *Staying with the Trouble: Making Kin in the Chthulucene*. Durham: Duke University Press.
Haseman, B. (2006). "A Manifesto for Performative Research." In *Media International Australia*, 118(1), pp. 98–106. https://doi.org/10.1177/1329878X0611800113
Jaque, A., Verzier, M. O. and Pietroiusti, L. (2020). "Editorial." In Jaque, A., Verzier, M. O., Pietroiusti, L. and Mazza, L. (eds.) *More-than-Human*. Rotterdam: Het Nieuwe Institut, Serpentine Galleries, Office for Political Innovation and Manifesta Foundation.
Jensen, P. (red.) (2017). *The Ethology of Domestic Animals: An Introductory Text*. 3rd ed. Wallingford, Oxfordshire, UK: CABI.
Jones, S. and Woglom, J. F. (2015). "Behind the Body-Filled Scenes: Methodologies at Work on the Body in Graphica." In Perry, M. and Medina, C. L. (eds.) *Methodologies of Embodiment: Inscribing Bodies in Qualitative Research*, pp. 116–137. New York: Routledge.

Kassabian, A. (2013). *Ubiquitous Listening: Affect, Attention, and Distributed Subjectivity*. Berkley: University of California Press.

Kember, S. and Zylinska, J. (2012). *Life after New Media: Mediation as a Vital Process*. Cambridge, MA: MIT Press.

Krause, B. L. (2012). *The Great Animal Orchestra: Finding the Origins of Music in the World's Wild Places*. New York, Boston, London: Little, Brown and Company.

Lochhead, J., Mendieta, E. and Smith, D. et al. (2021). "Introduction." In Lochhead (ed.) *Sound and Affect. Voice, Music, World*. Chicago and London: University of Chicago Press.

Lünenborg, M. and Maier, T. (2019). "Analyzing Affective media Practices by the Use of media Analyses." In Kahl, A. (ed.) *Analysing Affective Societies. Methods and Methodologies*, pp. 140–161. London and New York: Routledge (e-book).

Magnat, V. (2019). *The Performative Power of Vocality*. New York, NY: Routledge.

Manning, E. (2009). *Relationscapes: Movement, Art, Philosophy*. Cambridge, MA: MIT Press.

Manning, E. (2016). *The Minor Gesture*. Durham & London: Duke University Press.

Mazzei, L. (2009). "An Impossibly Full Voice." In Jackson, A. and Mazzei, L. (eds.) *Voice in Qualitative Inquiry: Challenging Conventional, Interpretive, and Critical Conceptions in Qualitative Research*, pp. 45–62. London: Routledge.

Menely, T. (2015). *The Animal Claim. Sensibility and the Creaturely Voice*. Chicago and London: University of Chicago Press.

Moisala, P. (2017). "'A People to Come' in Himalayan Village Music – A Deleuzian-Guattarian Study of Musical Performance." In Moisala, P., Leppänen, T., Tiainen, M. and Väätäinen, H. (eds.) *Musical Encounters with Deleuze and Guattari*, pp. 129–146. New York, NY: Bloomsbury Academic, an imprint of Bloomsbury Publishing Inc.

Moisala, P., Leppänen, T., Tiainen, M. and Väätäinen, H. (2017). "Introduction: Musical Encounters with Deleuze and Guattari." In Moisala, P., Leppänen, T., Tiainen, M. and Väätäinen, H. (eds.) *Musical Encounters with Deleuze and Guattari*, pp. 129–146. New York, NY: Bloomsbury Academic, an imprint of Bloomsbury Publishing Inc.

Mühlhoff, R. (2019). "Affective Resonance." In Slaby, J. and von Scheve, C. (eds.) *Affective Societies. Key Concepts*, pp. 188–199. London and New York: Routledge (e-book).

Mundy, R. (2018). *Animal Musicalities: Birds, Beasts, and Evolutionary Listening* [Electronic resource]. Middletown: Wesleyan University Press.

Neumark, N. (2017). *Voicetracks: Attuning to Voice in media and the Arts*. Cambridge, MA: The MIT Press.

Nyman, J. and Schuurman, N. (2016). "Introduction." In Nyman, J. and Schuurman, N. (eds.) *Affect, Space, and Animals*, pp. 1–9. London and New York: Routledge. (e-book).

Pedersen, H. (2011). "Release the Moths: Critical Animal Studies and the Posthumanist Impulse." In *Culture, Theory and Critique*, 52(1), pp. 65–81. https://doi.org/10.1080/14735784.2011.621663

Pitt, H. (2017). "An Apprenticeship in Plant Thinking." In Bastian, M., Jones, O., Moore, N. and Roe, E. (eds.) *Participatory Research in More-than-Human Worlds*, pp. 92–105. London and New York: Routledge (e-book).

Plec, E. (2013). "Perspectives of Human-Animal Communication. An Introduction." In Plec, E. (ed.) *Perspectives on Human-Animal Communication*. New York: Routledge.

Rosenberg, S. (2013). *Kurbits-ReBoot: svensk folksång i ny scenisk gestaltning*. Diss. Helsingfors: Sibelius-Akademin, Helsingfors.

Rosenberg, S. (2019). "Kulning nu." In *Fäbodlandskap och vallmusik*, pp. 160–171. Falun: Dalarnas fornminnes- och hembygdsförbund.

Rothenberg, D. (2001). “Introduction. Does Nature Understand Music?” In Rothenberg, D. and Ulvaeus, M. (eds.) *The Book of Music and Nature: An Anthology of Sounds, Words, Thoughts*, pp. 1–12. Middletown, CT: Wesleyan University Press (e-book).

Rothenberg, D. (2013). *Bug Music: How Insects Gave Us Rhythm and Noise*. New York: St. Martin’s Press.

Schötz, S. (2019). “Paralinguistic Information and Biological Codes in Intra- and Interspecific Vocal Communication: A Pilot Study of Humans and Domestic Cats.” In *Fonetik 2019*, Stockholm, Sweden. https://doi.org/10.5281/zenodo.3246003

Seeger, A. (2016). “Natural Species, Sounds, and Humans in Lowland South America. The Kĩsêdjê/Suyá, Their World, and the Nature of Their Musical Experience.” In Allen, A. S. and Dawe, K. (eds.) *Current Directions in Ecomusicology: Music, Culture, Nature*, pp. 89–98. New York, NY: Routledge.

Silvers, M. (2020). “Attending to the Nightingale: on a Multispecies Ethnomusicology”. In *Etnnomusicology*, 64(2), pp. 199–224. https://www.jstor.org/stable/10.5406/ethnomusicology.64.2.0199

Simonett, H. (2016). “Of Human and non-Human Birds. Indigenous Music Making and Sentient Ecology on Northwestern Mexico.” in Allen, A. S. and Dawe, K. (eds.) *Current Directions in Ecomusicology: Music, Culture, Nature*, pp. 99–108. New York, NY: Routledge.

Small, C. (1998). *Musicking: The Meanings of Performing and Listening*. Hanover, NH: University Press of New England.

Stark, H. (2017). *Feminist Theory after Deleuze*. London: Bloomsbury Academic.

Stewart, K. (2007). *Ordinary Affects*. Durham, NC: Duke University Press.

Stomberg, I. and Tiderman-Österberg, J. (2021).” Duets with Nature. How Natural Acoustics Affect the Experience of Performing Herding Music in Outdoor Settings.” In *AAWM Journal Music and Nature*, vol. 1. https://zenodo.org/record/5608379#.YY41kWDMJPY

Taylor, H. (2017). “Marginalised Voices. Zoömusicology through a Participatory Lens.” In Bastian, M., Jones, O., Moore, N. and Roe, E. (eds.) *Participatory Research in More-than-Human Worlds*, pp. 38–53. London and New York: Routledge (e-book).

Taylor, H. (2020). “How Musical Are Animals? Taking Stock of Zoömusicology’s Prospects.” In *Music Research Annual*, 1, pp. 1–35.

Thompson, M. (2013). “Three Screams.” In Thompson, M. and Biddle, I. (eds.) *Sound, Music, Affect: Theorizing Sonic Experience*, pp. 147–162. London: Bloomsbury Academic.

Thompson, M. and Biddle, I. (2013). “Introduction: Somewhere between the Signifying and the Sublime.” In Thompson, M. and Biddle, I. (eds.) *Sound, Music, Affect: Theorizing Sonic Experience*, pp. 1–24. London: Bloomsbury Academic.

Tiainen, M. (2017a). “Singing Non-Human-Centric Relational Futures – *The Algae Opera* as an Assemblage.” In Moisala, P., Leppänen, T., Tiainen, M. and Väätäinen, H. (eds.) *Musical Encounters with Deleuze and Guattari*, pp. 87–106. New York, NY: Bloomsbury Academic, an imprint of Bloomsbury Publishing Inc.

Tiainen, M. (2017b). “Sonic Technoecology: Voice and Non-Anthropocentric Survival in The Algae Opera.” In *Australian Feminist Studies*, 32(94), pp. 359–376. https://doi.org/10.1080/08164649.2017.1466651

Tiainen, M. (2018). “Sonic Performance and Feminist Posthumanities. Democracy of Resonance and Machinic Sounds.” In Åsberg, C. and Braidotti, R. (eds.) *A Feminist Companion to the Posthumanities*, pp. 103–115. Cham: Springer.

Tiderman-Österberg, J. (2019a). “Dit lockropen tog mig.” In Thorell, E. and Tiderman-Österberg, J. (eds.) *Fäbodlandskap och vallmusik*, pp. 190–205. Falun: Dalarnas fornminnes- och hembygdsförbund.

Tiderman-Österberg, J. (2019b). "Kom Kullorna! Kom Vallros." In Thorell, E. and Tiderman-Österberg, J. (eds.) *Fäbodlandskap och vallmusik*, pp. 172–189. Falun: Dalarnas fornminnes- och hembygdsförbund.

Tiderman-Österberg, J. (2019c). "Reflektioner och processer." in Thorell, E. and Tiderman-Österberg, J. (eds.) *Fäbodlandskap och vallmusik*, pp. 150–159. Falun: Dalarnas fornminnes- och hembygdsförbund.

Tiderman-Österberg, J. (2020)."Why Sweden's Ancient Tradition of Calling Home the Herds Is Women's Work," in *Smithsonian Magazine*. https://www.smithsonianmag.com/smithsonian-institution/why-swedens-ancient-tradition-calling-home-herds-womens-work-180975904/

Tiderman-Österberg, J. (forthcoming in 2023). "Nordic Fäbod Culture." In Birkehag, E. (ed.) *D for Daughter*. Leipzig: Spector Books.

Volgsten, U. (2019) "The Feeling of Music: Affect, Attunement, and Resonance." In Antenor Ferreira Correa (ed.) *Music, Speech, and Mind*. Associação Brasileira de Cognição e Artes Musicais - ABCM Music and Cognition Series.

Willett, C. (2014). *Interspecies Ethics. Critical Perspectives on Animals*. New York: Columbia University Press.

Wolfe, C. (2009). "Human, All Too Human: 'Animal Studies' and the Humanities." In *PMLA*, 124(2), pp. 564–565.

Yunkaporta, T. (2020). *Sand Talk: How Indigenous Thinking Can Save the World*. 1st ed. (e-book). San Francisco: HarperOne.

9 Speaking of atmospheres

More-than voice and voice of the more-than

Norie Neumark

In this chapter, I want to open a conversation between voice, place, and new materialism—through speaking of atmospheres and listening to them speak. To do this, I will engage with a number of artworks through voice theories, new materialism, and cultural geographical approaches to place. Attuning to atmosphere and voice together, I will ask how more-than-human voices help us apprehend atmospheres—and vice versa.

I will begin with atmospheres of legal institutions, listening to the artwork of Joel Stern (*I Am Sitting in a Courtroom*, 2017), which takes place in a law court, and the work of Lawrence Abu Hamdan (*Rubber Coated Steel*, 2016), exhibited in an art gallery. I will then turn to my own artwork with nonhuman animals, thinking about how emplacing the work in very different locations evoked different atmospheres. By attending to how atmospheres and voices speak and co-compose these artworks, I propose that we can apprehend anew voice, atmosphere and, indeed, new materialism.

Introduction—Speaking of atmospheres

Speaking of atmospheres evokes something palpable, something that moves us—inviting or repelling us. When atmospheres speak, they tincture our relations to a place and the people and things that co-compose it. We sense and respond to an atmosphere—of a room, say, or an artwork or a garden—but even as we react to its call, we also sense something more. A *more-than*-ness. A more-than-ness that reminds me of voice. Because like voice, atmospheres call out and connect. And, like voice—with voice—atmospheres affectively "traverse distinctions between peoples, things, and spaces," attuning us to ambiguities "between presence and absence, between subject and object/subject and between the definite and indefinite," as cultural geographer Ben Anderson analyses (Anderson 2009: 77–78).

Voice and atmosphere have much to say as they resonate with each other. Listening to voice can alert us to the complex workings and more-than-ness of atmospheres. And, in turn, attuning to atmospheres can expand our thinking with and through voice. With a new materialist ear, I will listen to how atmospheres' voicings beckon us to rethink the vital materiality and

DOI: 10.4324/9781003390992-13

multivocality of places, as well as the things and assemblages that inhabit and co-compose them. Inspired by the way that the Danish word for voice (*stemme*) also evokes the atmosphere (*stemming*)[1] of a place, I propose to explore the immanent entanglements of atmosphere and voice. I will ask how atmospheres are animated and expressed by voice and how voices are tinctured by atmospheres in the artworks I encounter.

New materialism ...

But before turning to the question of how atmospheres voice themselves, I need to briefly discuss how I understand new materialism and voice themselves, both of which I have elaborated elsewhere (Neumark 2017). As a Foucauldian toolbox of concepts, new materialism is agile, indeterminate and multiple. Working across culture, politics, ethics and aesthetics, new materialism has been able to open new questionings of anthropocentrism, human subjectivity, human exceptionalism and the nature/culture divide (Connolly 2013: 399). While I should note that many of new materialism's concerns around the human subject are shared by theorists of posthumanism, I prefer to work with new materialism, because it speaks *transversally* to artists, cultural and political theorists, and philosophers alike. This includes, for example, key writers in the arts such as Katve-Kaisa Kontturi, Iris van der Tuin and Estelle Barrett and Barbara Bolt for whom materiality is always palpable and agential—moving and dynamic, speaking and remembering (Dolphijn and van der Tuin 2012; Barrett and Bolt 2013; Kontturi 2018).

In *New Materialism: Interviews and Cartographies* (2012), Rick Dolphijn and Iris van der Tuin avoid setting up new materialism as a singular theory. Rather, as they explore enactments, movements, entanglements, and intra-actions with new materialism, they map the various ways of thinking—rereadings and rewritings—that produced it and that it produces. Van der Tuin and Dolphijn foreground new materialism's transversality across "feminist theory, science and technology studies, and media and cultural studies" (Dolphijn and van der Tuin 2012: 100). To these, I would add art practices which themselves are material ways of thinking, doing and producing (Ingold 2011). Inspired by Karen Barad's *intra-active* theorizing, I propose new materialist artful and collective ways of thinking, these "living and breathing reconfigurings of the world," as "responsible and responsive to the world's patternings and murmurings" (2012: 207). Barad's attunement to the world's deep murmurings, its *hum* (Barad 2012: 218, see also Barad 2015), returns me to voice.

Voice ...

Understood through new materialism, voice is more-than—more-than-human, more-than-sound, more-than-music.[2] From plants, to animals, to things, to places—to assemblages themselves—voices forcefully invite our listening as they evoke, convoke, provoke. With an agency and vitality, to which new materialism attunes us, voice calls out, calling upon others to listen and to respond. This is not a call constructed by the listener, human or otherwise, but a call from, a call upon, a call to, a call with—a call that affects and effects.[3] In her book *Vibrant Matter*, which has been influential for new materialist thinking, Jane Bennett regularly evokes voice both literally and metaphorically to convey the expressivity of things and assemblages. "Thus spoke the grid," she says in her discussion of the New York blackout of 2003 (Bennett 2010: 36). Like Karen Barad, Bennett recognizes this speaking as something not just humans do; and she proposes its more-than-human expressivity as something to which humans need to attune (Bennett 2010: 104). For Dolphijn and van der Tuin (rereading philosopher Vicki Kirby), "matter appears as some-thing that is not only spoken about or spoken with, but rather as itself something simply speaking" (Dolphijn and van der Tuin 2012: 108). For Bennett and other new materialist thinkers (and practitioners), listening to these voices is a way of attending to a thing—not to its experience from its point of view, which we cannot penetrate though we can sense it—but to its performativity in an assemblage. Thus new materialism can focus our listening to the voices of things within assemblages and, further, the voices of assemblages themselves. In short, new materialism invites an apprehension of voice both as a co-composing part of assemblages and as shaped by and speaking their atmospheres.

More-than ...

The assemblages within the atmospheres that I listen to in this chapter are co-composed with more-than-human voices. *More-than* is about collectivity; it both includes and questions what follows it—doing and undoing at the same time. *More-than-human*, then, includes as well as disrupts 'human.' When process philosopher Erin Manning works with *more-than*, she focuses on relationalities, movements, multiplicities, forces and ecologies. Writing of the *more than one* body, for instance, she evokes "the body as the complexity of imminent collectivity, the body as *resonant materiality* .... The body, always more than one, replete with the force of life ..." (italics are mine; Manning 2010: 118). Just as for Manning bodies are always-already an assemblage in motion, in process, in practice, emplaced in an environment ... so too are the human and more-than-human voices that co-compose an atmosphere. In the rest of this chapter, I will attend to the ways both more-than-human voices and the more-than-human of the human voice itself help us think atmospheres—and vice versa. But first, let me expand on atmospheres.

Part 1—Foregrounding atmospheres

For Gernot Böhme, a key thinker in this area, atmosphere speaks to the life of an environment in a sensuous, performative and entangled way—fundamentally entangling subject and object. Atmospheres, for Böhme, are spaces *tinctured* through persons, who sense through their bodily presence, and through things, which sensitively articulate their presence through various qualities which he calls *ecstasies and which include voice or sound as well as color or smell* (Böhme 1993: 116, 122; 2013; 2017: 49 and passim). To the various makers of atmospheres that Böhme signals, including set designers, garden makers and town planners, I would add voice and sound artists and installation makers.

Another important scholar who approaches atmospheres transversally is Ben Anderson, whose work chimes more directly with new materialism. In what he calls "affective atmospheres," Anderson entangles the phenomenological with materialist approaches to atmospheres. Through materialism Anderson focuses on the spatiality of atmospheres—surrounding and enveloping but also turbulent, indeterminate, ambiguous. Through phenomenology, he is alerted to the affective qualities that may emanate from bodies but also exceed them—"a kind of indeterminate affective 'excess' through which intensive space-times can be created" (Anderson 2009: 80). Anderson argues for the ambiguities of the atmosphere and proposes that it disturbs an unhelpful distinction between affect and emotion. For Anderson, with Böhme, "to attend to affective atmospheres is to learn to be affected by the ambiguities of affect/emotion, by that which is determinate and indeterminate, present and absent, singular and vague" (2009: 80). This recognition of the presence of absence in atmospheres is important and helps us attune to the voice of silence, to which I'll return later. What is also especially inviting in Anderson's and his collaborator James Ash's work is the way they bring affect into play with relationality: atmospheres envelop, they propose, as they are "neither wholly separate from the relations that form them, nor wholly determined by those relations" (Anderson and Ash 2015: 42–43). Again, this sounds like voice, as it resonates with voice's own paradoxes (Neumark 2010).

Anderson and Ash apprehend atmospheres as "part of the 'ubiquitous backdrop' of life and thought, while at the same time exerting some kind of force" (Anderson and Ash 2015: 35) Critic and writer of atmospheres, Kathleen Stewart, similarly perceives the force of atmospheres:

> An atmosphere is not an inert context but a force field in which people find themselves. It is not an effect of other forces but a lived affect – a capacity to affect and to be affected that pushes a present into a composition, an expressivity, the sense of potentiality and event.
>
> (Stewart 2011: 452)

Atmospheres—as Stewart senses them, lives them, and writes them—are simultaneously both poetically affective and commonplace. Commonplace. Common … place. Acutely attuned to their compositions of place, Stewart's writing listens to the voices of these precarious and complex atmospheres—every day and ubiquitous—bringing them from the background to the foreground (Stewart 2012: 519).

From background to foreground. This is the work of speaking of an atmosphere. When I began writing this piece, I serendipitously became involved in a Liquid Architecture project and reading group in Melbourne, Australia, *Why Listen to Plants*. Reading plant philosopher Michael Marder's grounded thinking with plants helped me appreciate this atmospheric background to foreground movement. Marder reminds us how usually plants "form the inconspicuous backdrop of our lives—especially within the context of 'urban landscaping'—much like the melodies and songs that unobtrusively create the desired ambience in cafes and restaurants" (Marder 2013: 3–4). Rather, attuning to the vitality, voices and thinking of plants, Marder brings them to the fore. His 'plant thinking' offers a way to "encounter plants" by listening to them and by thinking about and with them (Marder 2013: 3). Sensing how plants enliven atmosphere with their own vitality, plant thinking invigorates thinking about atmospheres and new materialism with it.

I responded in particular to Marder's plant thinking about the figure of ambience, because atmosphere is sometimes reduced to a commodified version of ambiance—to be consumed as part of the 'experience economy' with its promise of authenticity.[4] It is worth a digression here to essayist and artist Seth Kim-Cohen's critique of the ambient, particularly ambient sound art, to help further elaborate how literally and metaphorically atmospheres voice a place differently than—more-than—the commonly figured ambiance. For Kim-Cohen, ambient sound art is a contentious and worrying concept and practice, because it "downplays other situational and power relations" (Kim-Cohen 2016: 10). Kim-Cohen rejects the way ambient art brackets politics and context and site—and here I would add place (Kim-Cohen 2016: 33). For him, the ambient ignores the *with-ness*, the relationality that is crucial to life and art. And, I would add, to atmosphere and voice. He works with Heidegger's figure of *mitsein*—being with—to think about sound as having its identity 'with' other things in the world—as compared to ambient's assertion of an immanent 'thing-in-itself' (Kim-Cohen 2016: 57–59). For me, Kim-Cohen's sense of context speaks of atmosphere and what it voices. That is, what Kim-Cohen says about with-ness and foregrounding helps us understand atmosphere not as ambient ambiance but as a social, cultural, political, and aesthetic force field—as Michael Marder would have it, a *milieu*—a middle—by definition a relational space (Marder 2013: 63–64).

This relational space has been apprehended by sound theorist Douglas Kahn as *transperception*. Kahn offers as an example Henry David Thoreau[5] who listened to distant church bells tolling through, and with, the 'vibrating' atmosphere. Thoreau heard the bells as having "conversed with every leaf and needle of the woods. It is by no means the sound of the bell as heard near at hand … but its vibrating echoes, that portion of the sound which the elements take up and modulate—a sound which is very much modified, sifted, and refined before it reaches my ear" (cited in Kahn 2013: 170). Thoreau as plant thinker …

While Kahn and Thoreau may be more directly concerned here with sound rather than voice, the figure of *conversing* speaks of voice to me, as the bells and the atmosphere and the woods *converse* on their way to Thoreau's listening ears. Indeed, Kahn goes on to note that for Thoreau the echo of his voice in the woods was "not merely a repetition of my voice, but it is in some measure the voice of the wood" (cited in Kahn 2013: 171). Attuned to voice and atmosphere through the figure of transperception, new materialism might rethink vibration (all too often approached reductively and singularly in sound studies) together with environment, mediation and listening—the atmosphere voicing itself (Neumark 2017: 65).

Finally, let us listen to how Indigenous Australians' sense of country speaks importantly to understandings of place, of relationality, of atmosphere. Asked about plant thinking for the Liquid Architecture project, Dr Glenda Harward-Nalder responded:

> First Nation Peoples listen to plants and plants listen to us. Ours is a reciprocal relationship. In First Nation ways of relating, the people belong to the land in community with all living things. In our language places are named for the plants or foods that grow there, e.g., **gulburbin** (place of the dugong grass) or, **ngarangarawayi** (long winding creek where swamp heather grows abundantly) or for their special geographical features ... In our culture, we belong to country, where all things are interrelated, and nature and culture co-evolve.
>
> (Harward-Nalder 2017: 1)

This relationship of enveloping reciprocity has much to teach new materialism and helps attune us anew to atmospheres.

Part 2—Speaking of institutional atmospheres

Listening now with expanded, new materialist approaches to atmospheres and how they voice themselves, I want to turn to two artworks that speak institutional atmospheres. Despite their common hushed, reverent quiet, law courts and art galleries are very different institutions whose silences evoke very different atmospheres and affects in artworks sited there. But before we can listen to the quiet of institutional atmospheres, we need to attend to the particular politics and ethics of silence (Ihde 2007: 180 and passim; Voegelin 2010: 83). Silence is always-already richly alive with voice—voices of memory and voices of imaginings. Resonating across culture, history and art, silence vibrates with a present/absent fullness—voices waiting, calling out to be listened to. And it speaks of voices we are not, yet, attuned to hearing. And in different institutions, silences speak their different atmospheres. This means artworks engaging with legal, institutional concerns call to their publics very differently in these different locations, composing very different atmospheres.

Located in the Melbourne Federal Court Building, Liquid Architecture and the Melbourne Law School's *Acoustic Justice Project* (July 2017) understood the atmospherics of a court room's silence, explaining:

> Sound is essential to the administration of justice, an inalienable part of our legal worlds. But not just any sound. Listen carefully. What do you hear? To begin with, silence: a powerful quiet. Mobiles are switched off. Conversations whispered. Soundproofing isolates against the volume of daily life. The judicial soundscape depends on and entrenches an association between silence and civility, noise and disorder, with an exceptionally long pedigree in the West and elsewhere. In the courtroom, silence is figured as the proper condition out of which legal discourse emerges.
>
> (Liquid Architecture and Melbourne Law School 2017)

One of the works in the *Acoustic Justice* event that made the atmosphere most palpable for me was Joel Stern's *I Am Sitting In A Courtroom* (2017). What he did in the work was re-place Alvin Lucier's famous sitting in a room of musical composition with sitting in a courtroom. Stern's work has been haunting me for a while as I have wondered where and how I could think/write about it. I went to the performance partly drawn by the chance to go inside a courtroom where civil and criminal trials take place. I found myself sitting in the back of the public gallery—a legal gallery, not art gallery—even though art was taking place here. When Stern's work began, I was immediately immersed in a strange atmosphere (legal and art entangled) as the 'court clerk' called Joel Stern to the stand—and called me in, called me to order and recalled Alvin Lucier to my memory.

The atmosphere was replete with the voicing of an uncanny place—a gallery but not an art gallery, a court room but with nothing on trial except perhaps experimental sound art—as the performance inhabited the space, playing, playing with, and disturbing the protocols of place. The heavy, stale air of the court room deadened the atmosphere and demanded the silence that speaks of an institutional, legal setting. Yet strangely, enlivened by (art) gallery atmosphere, the people around me chattered and fidgeted, took photos and video with their phones, felt like an art crowd—haunting the atmosphere with very different proceedings, very different 'audiences.' As the performance began, we were stilled and called into this heterogeneous space, listening attentively, as an art crowd, to every sound as Stern poured water, and placed his glass on the table. The gesture of a performer or a lawyer or a witness ... a call to witness for us and to performance for him. And then came the hollowed sound of the microphone in the hallowed room. This unfamiliar but palpable atmosphere stirred me to watch and listen.

Riffing off Alvin Lucier's famous piece, *I am sitting in a room*, Stern's work also called for listening anew to speech, but this time foregrounding not resonant frequencies of a voice in a room but the hearing of language. Perhaps it was hearing on trial—a hearing in the legal sense and a hearing in the sensory sense. Unlike Lucier's composition with tape recording, the repetition in this case happened through collaboration with an actual court stenographer who typed, with stenographer's specific renderings, Stern's repetition of his text:

> I am sitting in a courtroom the same as the one you are in now. I am having the sound of my speaking voice transcribed and I am going to read it back to you again and again until this tacit operation of the courtroom reinforces itself so that the coherence of my speech, with perhaps the exception of isolated words, is destroyed. What you will hear, then, is this tacit operation of the courtroom articulated by speech and text. I regard this activity not so much as a demonstration of an institutional fact, but more as a way to draw out any irregularities of this hearing.
>
> (Stern 2017)

He read the transcription and she transcribed his reading, each time both becoming more humorously incoherent and precarious. The transcriber looked at him, he read her screen. The atmosphere was tinctured with uncanny humor and zany affect—as the court transcriber spoke to the artist and he asked if she'd transcribed that. "Stop stop stop." She inserted some comments. Laughter in the gallery. As the music began, Stern signaled the end by reading her writing ("open brackets gentle music plays"). Eventually, the music (and dancing) take over "(now annoying music)" "It is hard work for me too." Lots of brackets, but no bracketing of life, the world, institutions and their atmospheres—both art and legal.

For Stern, the work composed a 'feedback loop' between himself and the stenographer: a loop that created an experimental agitation of language—text transformed through disassembling. As he spoke the text as she typed, the peculiarities of stenographic 'language' colored his speech (Stern 2018). While Lucier's work had played with his speech impediment, transforming it into music which erased his stutter, Stern brought stuttering back as his voiced text disintegrated into a grammatical mess ... a degradation in the language ... The atmosphere in the room stuttered with it—between the artful and the legal—assembling and disassembling both.

In an interview with me, Stern explained that his work was interested in "the way speech acts in a courtroom." For me, the work brought to the fore how the atmosphere of a court room acts on speech and speech acts on atmosphere, by infecting it with an art gallery atmosphere through its art audience and art history reference. Stern wanted to make material what is lost and what is added in the translations of transcribing in a court room. To do this, he worked with Lucier's piece as a

> methodology for allowing something which is in the background of a scene or a situation to be intensified progressively up to the point

> where it completely takes over and becomes dominant. His work is about this background becoming the foreground or this atmosphere which is tacit becoming overwhelming and dominant .... The movement of something from the background to the foreground when it is structurally transparent tells you a lot about how that thing operates.
>
> (Stern 2018)

And it tells you about the atmosphere: "Drawing attention to the tacit operations of the court—that is quite close to ... atmosphere" (Stern 2018). Discussing tacit operations and atmosphere with me, Stern referred to Seth Kim-Cohen's essay *Against Ambience*, noting the way in which, with ambience, we sense something is present but don't know how it operates, which tends to naturalize its power. Like Kim-Cohen, Stern prefers to work against ambience to de-naturalize power. And this is what happened for me with Stern's work as it foregrounded and dismembered the atmosphere of a courtroom at the same time, and in so doing troubled the power that reverberates there. "I was in a way holding court." Holding the court to account, with its atmosphere of serious legitimacy, of institutional power over speech. If the silences of a courtroom (and an art gallery) evoke major atmospheres of solemnity and reverence, this performance evoked minor emotions—a disconcerting humor, not quite slapstick, not quite uncanny—zany.

For cultural theorist Sianne Ngai, in her discussion of marginal or minor aesthetics (2015), the zany is about very physical, intense performing, but not the sort of performing that achieves something or is useful. She sees it as an aesthetic of nonstop action, repetitive action. ("Stop. Stop. Stop.") It is an aesthetics that goes from sitcoms to Atari video games to avant-garde practice like Dada cabaret. Zaniness happens when someone experiences too much happening at once, too much coming at them at once. (Stern: "Now I'm totally lost.") Ngai sees zany as an "aesthetic of incessant doing." For me, Stern's disturbed repetitions make me sense the courtroom as "a theatre of justice" by "atmospherically re-composing" the courtroom, into something on the edge of zany.[6]

Stern's work speaks of the political and ethical dimensions of atmospheres. And atmospheres in artworks like Stern's speak of politics, the *p*olitics of minor gestures (Manning 2016; Miranda and Neumark 2018). For Erin Manning, artists' minor gestures are fragile and undetermined (Manning 2016: 1–7), stirring up the ground in ways that activate and make possible larger political shifts. They ask,

> how art can make a difference, opening up the existing fields of relation toward new forms of perception, accountability, experience, and collectivity. This aspect of the art of participation cannot be thought separately from the political, even though the work's political force is not necessarily in its content.
>
> (Manning 2016: 56)

There is an uncertainty and complexity to the encounters, 'co-compositions,' and engagements that minor gestures make possible (Manning 2016: *passim*, esp. 3, 7–8). In Jacques Rancière's sense, they reframe common sense—what he calls the 'common sensorium,' through the "distribution of the sensible, or the system of divisions and boundaries that define, among other things, what is visible and audible within a particular aesthetico-political regime" (Rancière 2004, Rockhill 2004: 1). And as Stern's work does for me, they disrupt and reframe the legal institutional atmosphere—an atmosphere whose power we would take for granted as natural—re-composing it and moving it to the fore as what is usually audible and visible is disrupted at the same time as it is made more palpable. The work attunes me to the way a courtroom is always-already pervaded by a heavy atmosphere, rent through with power relations. These were the 'tacit' power relations "vibrating in the background" that Stern brought into the foreground. The political atmosphere of a courtroom made his work stutter. In its surprising and zany way, Sterne's work voices the politics of atmosphere.

Which brings me to another disruption of courtroom and gallery, this time in an art gallery. And this time, no laughter. Here Joel Stern, along with collaborator James Parker, curated the *Eavesdropping* exhibition in a major Melbourne art gallery, bringing to a wider art public works that play with legal institutional atmospheres and silences. A particularly moving work was Lawrence Abu Hamdan's video, *Rubber coated steel* (2016). The atmosphere of the legal proceeding in the video itself and in the art gallery where I encountered it was very different from Stern's own artwork. Moving through the *Eavesdropping* art exhibition, I was acutely and palpably attuned to the perceptions, potentials and responsibilities of being an 'ear witness,' 'just as the curators invited. And even though the political 'issues' in the exhibition were familiar to me, in this disturbing atmosphere I was provoked to sense anew the importance and value of this artful listening—the artists' and my own—as I was aesthetically opened to unpredictable and deeper feelings and understandings. When I finally entered the cul-de-sac where Abu Hamdan's work was sited—dark and separate though other sounds did porously leak in—I was quickly immersed in an atmosphere that was dark in every sense.

Rubber coated steel is part of Abu Hamdan's wider work which researches the aesthetic possibilities of *forensic architecture*. As its founding director Eyal Weizman explains, forensic architecture is a political, cultural and artistic approach based on the potentials of *forensis*, which

> bring[s] new material and aesthetic sensibilities to bear upon the legal and political implications of state violence, armed conflict, and climate change. [It] seeks to perform across a multiplicity of forums, political and juridical, institutional and informal.
>
> (Weizman 2014: 9)

The transposing of forensics from traditional crime investigation to other forums was also crucial for voice practitioner-scholar, Konstantinos Thomaidis, with his interdisciplinary analysis of the potentials of forensics of voice in performance studies (Thomaidis 2018). Apprehending voice as "unruly and processual" (233), Thomaidis foregrounds the potentials of active, aesthetic listening to voice in performance—listening for clues and evidence in the voice—while at the same time recognizing the complexities and complicities of such analysis in practice. Voice as process, listening as process—these are crucial to the way Thomaidis brings forensics to the analysis of performance. He also points out the uncomfortable position of a forensic earwitness to a performance, a listener who can never be neutrally outside the scene because they are always part of the process:

> We, as spectators, are not a community uncomplicatedly united in the aftermath ofatrocity; 'we' as audiences, unite in the face of adversity that we (may have) brought upon ourselves. As if in the processes of forensically unpacking performed vocal evidence, our subject position hauntingly shifted from that of the earwitness to the revelation that 'we' might be the perpetrator—or in some measure, his unknowing accomplices.
> (Thomaids 2018: 233)

Haunting, indeed—haunting for me that while Thomaidis is analyzing quite a different performance, somehow it is as if he is listening to the same one that I am. It is as if he is witnessing the ability of Abu Hamdan's performance to provoke an uneasy sense that 'we' as earwitnesses might be the perpetrators. Because this is the very shift that Abu Hamdan achieves with his own performance of forensic evidence in an uncanny play of voice and atmosphere.

Let me return now to Hamdan's work to elaborate. The video *Rubber coated steel* was a silent enactment, restaging an investigative report's findings as a tribunal. Under investigation was the question of whether in May 2014 Israeli soldiers in the occupied West Bank (Palestine) used real bullets in guns specially fitted with a "rubber bullet adaptor" to suppress their live bullets' sound—in an attempt to fool vulnerable Palestinian youths into thinking the bullets were rubber. The report and video tribunal were concerned with whether by thus making the live shot sound like rubber bullets, the soldiers were able to kill with impunity two Palestinian teenagers, Nadeem Nawara and Mohamad Abu Daher.

Unlike a shooting range where you hear the bullets' deafening voice, you don't actually hear the bullets in this work, in this room–which recalls the way rubber bullets dampen their own sound. And so, what we experience in Abu Hamdan's video is a noisy event but an atmosphere of silence—which recalls to mind and feeling the imposed silences of Israeli politics and practices. Although it tells of a legal investigation, the video places us not in a familiar courtroom but looking into a shooting range, in which images, like targets, are brought literally and figuratively to the fore on target retriever machines. Their CLUNKs puncture the room's silence. (Not a complete silence as sounds of other nearby artworks leak in—reminding us of the art context and place of this work.) At the bottom of the screen, we read the transcript of the tribunal and 'listen' to its sound: "Audience winces at loud sounds." In the video, the (mostly) silent soundtrack leaves the bullets' sounds to our disquieted imagining—rendering a disquieting atmosphere. Nor do we hear the voice of the judge (except through his muted gavel), prosecution or defense—nor the families of the murdered or other youths. Reading their testimonies in subtitles, I react along with the tribunal's audience. I oscillate between being inside and outside the tribunal, jury and observer, earwitness and perpetrator. I am both the video's audience and the tribunal's audience, though unlike the latter I only imagine the sound of the bullets—the sound with which the silence is replete—as the image-print of each type of bullet, among other images, is pulled to the foreground by target retrievers. CLUNK. I can only imagine the difference between the sounds of a rubber coated steel bullet and an uncoated steel bullet. In any case, it is unlikely that I could tell the difference between live and rubber bullets in the way that the "true witnesses," the Palestinian youths, have learned to—as a matter of life and death.

The *mise-en-abyme* in this work takes me into an abyss of dread, its heavy atmosphere troubles me. Mesmerized by Abu Hamdan's video, I slowly realize in my growing dismay that I am standing in the position of the shooter in the gun range bay that I am watching on the screen. I am forcefully and uncannily positioned inside the work and feel deeply implicated in the distressing tribunal that is unfolding in the darkness of the video's room within the dark room in which I am standing. A dark atmosphere of deafening silences around Israeli violence to the Palestinians overwhelms me. Silence reverberates here. The familiar bullets' voices of shooting range are de-familiarized—here instead the atmosphere reverberates *silently* with mediated memories of those bullets' voices—memories uncannily silenced … as if by the rubber bullets themselves, co-composing the silence of this atmosphere that engulfs me. It is a silence alive with more-than-human voices.[7]

This noisy silence of Abu Hamden's work spoke, too, of temporality and the temporality of atmosphere—the daily time, marked by gunfire and the memory of gunfire, of Palestinians living under Israeli force. For theorist Don Ihde, "silence is the sound of time passing" (Ihde 2007: 111). And it spoke of ethics. I listen reminded of Ihde's *ethical* sense of the silence, which listening, and then speaking, must respect (Ihde 2007: 180 and passim). A political, ethical urgency composes the atmosphere of this artwork.

Part 3—Making/voicing an atmosphere with worms

Weighed down by these memories, I turn now to quite a different work—my own project with my long-time collaborator, artist Maria Miranda. This work, *Waiting*, was part of a project animated by *with-ness*—an attempt as artists to work *with* animals rather than use them or display them. To do this, we wanted to attune to worms and to the atmosphere in their world, and ours, through first listening to their voices and amplifying them through our own bodies. We composed the work with the worms in very different places, with different atmospheres—a railway station waiting room, an art gallery, and our own back yard. The atmospheres were made different by the work and the work was made different by the atmospheres. Each place spoke the work, and the work spoke the places.

Back yard ... The project began when we moved from a flat in Sydney to a house in Naarm/Melbourne. Suddenly, we had a backyard, veggie patches, and a compost bin—a very Melbourne inner suburban way of life. We hadn't thought about worms before, but we were quickly enchanted. The more we fed them and watched them, the more fascinated we became. We discovered shared passions, like the love of coffee and a vegetarian diet. We were entranced by the way these small, humble creatures have such a powerful effect on our own ability to live healthy, happy lives. They literally make the topsoil that grows our food. We were further drawn to work with worms when we sensed an affinity between our commitment to recycling and their composting/transformational skills. We loved that they still retain much mystery, at the same time as being a common—though often unnoticed—part of everyday life. Our connections to earthworms—so vital to everyday life and to the existence of humans, nonhumans and the environment—are normally overlooked in the background, even despised. And we responded to their place in the atmosphere of groundedness—their very literal groundedness—as they are in, of, and make the place they and we are in together.

Animated by the worms, our backyard was tinctured by an atmosphere of enchanted *companionship*. Ecofeminist Donna Haraway digs into the roots of *companion* and finds *cum panis*–with bread (Haraway 2008: 17). This reminder of how breaking bread and eating together lie at the heart of companionship made me notice how our worms, companion-critters, eat *with* us. As I put my breakfast makings and leavings into the worm café and compost bin, I realized that it is not just that we were feeding the worms but that they were actually eating what we eat—we were all eating the same food, drinking the same drinks (yes, an incredible amount of coffee and tea and quite a lot of toast, speaking of *panis*). As I drank my/our morning coffee and munched my/our toast, I remembered, too, that Charles Darwin admired worms' companionability, their sociability: "They perhaps have a trace of social feeling, for they are not disturbed by crawling over each other's bodies, and they sometimes lie in contact" (Darwin 2007: 15). Or, as Donna Haraway might put it with more of a new materialist twist, worm bodies are "entangled meaningful bodies" (Haraway 2008: 26).

Called to make a work with our companion worms, we were moved, like many artists, by a crucial ecological and new materialist question of how to work with nonhuman animals ethically and politically as well as aesthetically ... how to work in ways that don't play on romantic sentimentality nor exploit their vulnerability. How, in other words, to work *with* them, attending to their intelligence and creativity rather than simply putting them on display? We were inspired by theorist Vinciane Despret who comes at these issues through the figure of *with-ness*—describing possibilities of human-animal relations as a mutual *attunement*, a passionate, bodily *with-ness*. For Despret, these relations of attunement depend on the availability of the bodies to each other, to practices that transform both, to affects that move them. She proposes that in relations of attunement, it is affect that articulates bodies, that creates and is created: that is, attunement is how affect shapes the bodies in relationships, activating them (Despret 2004). For Bruno Latour, following Despret's argument, attunement happens through "body talk … the many ways in which the body is engaged in accounts about what it does" (Latour 2004: 206). While Latour is probably focused on body here, I listen to *talk* speaking just as loudly.

Vinciane Despret's *with-ness* understands the encounters between humans and animals as a blurring of knowing and known bodies, because of the way bodies of both enact 'embodied empathy' in these encounters (Despret 2013: 69). "Empathy, in this case, is not feeling what the other feels, it is rather making the body available for the response of another being" (Despret 2013: 70). Embodied empathy is not about humans feeling what nonhuman animals feel (or vice versa), but feeling with them, thinking with them, an "embodied communication" (Despret 2013: 71). This is what we felt with the worms, and this embodied empathy enlivened our backyard and the work we did there.

And so, Maria and I decided to work with worms as artists as a way to think through making, as a way to engage with these questions of how to attune to nonhuman animals—how to work *with* them, how to 'follow' them, as Kaisa Kontturi puts it (Kontturi 2018).[8] We began the project *Waiting* as an ongoing durational art project in May 2017. With this project, we wanted to make sound and video compositions that were open provocations about what is creativity, what is thinking, and what are the possibilities for human-nonhuman animal relations. Looking back, we were responding to the *enchanted* atmosphere of our backyard, an atmosphere enlivened by our relationship with the worms, which we hoped would, in turn, enliven the artwork (Bennett 2001). Speaking of enchantment, Jane Bennett offers an important re-apprehension of enchantment and wonder, as when/where we are "caught up and carried away" (Bennett 2001: 5). Through her 'aesthetic-affective' understanding of the ethics of enchantment, Bennett proposes that to *sense* the enchantment and wonder of the modern world is vital to reanimate the engagement necessary to motivate generous ethical behavior. Just as the worms enchanted the atmosphere of our backyard, we hoped our work would resonate aesthetically and ethically with this wonder.

The waiting room of the railway station … In our backyard, we developed methodologies of waiting and listening as we made the videos and took photos of the worms in the worm cafes and compost bins. Taking the photos and videos went on for months, since the worms did not always want to come out into vision. We learned to wait to watch. Meanwhile, we thought about how we could listen in a way that also disrupted the worms and their place as little as possible. We decided to carefully place a contact microphone in the worm, *Schhlrrsshh. Chlwllchch. Fthststststsss.* As we listened to—*with*—their moving voices, as we followed their voices, we attuned to their voicing moving through headphones: we felt their voices affectively move through our bodies as our bodies *amplified* their voices. By videoing ourselves doing this—listening and amplifying their voices—we chose to put our bodies on the line rather than theirs. We also wanted the video to present our direct encounter. Rather than abstracting the worms' voices through *sonification*,[9] we preferred a sonic expression of more-than-human voice, in an intra-active relationship.

To share our experiences with worms, we had proposed a site-specific sound and video installation, *Waiting*, for the MoreArt Festival in Melbourne, a public art festival run by the municipal council, at the end of 2017. We were commissioned to make and install the work in the suburban Moreland Railway Station Waiting Room. Besides the resonance of the waiting room site with our waiting project, the theme of the MoreArt festival, 'thresholds,' spoke to us. At certain unknown points—through proximity, through time, through ongoing care and through waiting—a threshold was reached and our relationship with the worms would change. We would sense, for instance, their agency and their temporality when they 'refused' to come up for a photo shoot and we had to wait—to adjust our art-making timeframe to theirs. At these threshold points, something happened that shifted our experience of the world. It was not so much an endpoint as a beginning. This is what the installation aimed to explore for us and to share with the audience—a threshold moment where the background atmosphere would come to the fore, to be experienced.

The waiting room at Moreland railway station is a large, cold room with no windows and very little light. It is usually locked up with barred gates, which allows a certain amount of filtered light through and allows people to peer between the bars into the room. It was a cold, uninviting atmosphere for the most part, yet paradoxically we also experienced it as full of promise, too: promise of potential connections with an audience. As with most of our site-responsive installations, we worked with the space to make the work. We first spent time in the space to get a feeling for it. One of the outstanding things we noticed about the waiting room was that people waiting for a train spend on average 5–10 minutes sitting or pacing the platform. The audience would be spasmodic and follow the train timetable. Very quickly we also realized that this space was different from a gallery, with the usual art audience. And the oncoming and departing trains overwhelmed the voices of the soundtrack every few minutes. The waiting room was a public space, and people could look through the bars at the work when we weren't there and be welcomed in to wait with us when we were. Early on, we began thinking about whether to bring the worms into the waiting room—how would this atmosphere affect them and how would their presence affect it?

Once we began installing in the space, we sensed that the worms would be too vulnerable there and the timetable-based waiting at a train station did not resonate with the process-based waiting that worms invite. We worried that people would want to open the lid of the worms' boxes to see them in *action*. So we decided to leave them safely at home and instead to bring in their photo portraits and their contributions ... the castings that they make. These warmed the atmosphere and enlivened it without putting the worms at risk. In this way, the atmosphere was also tinctured with an art gallery edge by multiple, companionable worm portraits—given the sociable worms weren't given to individual posing, each portrait was multiple.

The gallery ... From the messy, noisy, busy atmosphere of a railway waiting room to the quiet emptiness of a small space in a white cube art gallery. The invitation to site the work in a small artist-run space raised new questions for us about how to install the work, how it would respond to this new atmosphere and what it would bring to that atmosphere. How could we bring forth the atmosphere of life and liveliness that we felt at the railway waiting room? With this version of the project, *Worms & Worms & Worms*, we decided not to fight the starkness of this art space. Instead, we worked with unexpected shifts in scale to foreground the starkness and smallness and shape of the room and to foreground the worms. We filled the back wall with a large video of the worms. On the floor to the side and on the wall on the other side (on a tiny projector and iPad) we placed very small videos of ourselves—listening to the worms with headphones and amplifying their voices. Serendipitously, the sound system for the whole room didn't work so the only access for the audience to the sound was to listen through headphones on the iPad on the wall. This created an atmosphere of intimacy of listening—evoking the intimacy of our relationship with our worms—as the listeners watched us, small, in front of them, ourselves listening to the worms with headphones and then turned to watch the worms, writ large, on the back wall. There was a gentleness and slowness to the worms' movements that warmed the cold gallery atmosphere and invited the audience to feel their presence in their absence. With this shift in scale and the intimate voice of the worms/us amplified, the backyard's atmosphere seeped into the gallery.

Conclusion

The atmospheres in this chapter's artworks have each expressed a *more-than*-ness of voice and voicing. Sometimes, their atmospheres called to me in an enchanted and empathetic way while at other times they disturbed and troubled me with zany humor or vociferous silence. With all their different voices and voicings, each of these works has spoken of the complexities and common/place-ness of atmosphere. By attending to them with a new materialist ear, I have tried to listen to their voices and atmospheres anew. New materialism's attention to material agencies and lively assemblages has foregrounded in fresh ways the cultural and political contexts of these voicing and speaking atmospheres. It has insisted on the aesthetics and ethics of emplaced voice and of agential atmosphere.

With atmosphere opening up our sense of voice and place, and new materialism alerting us to the social, political, and cultural ethics and aesthetics that tincture them, we can apprehend how atmosphere is inseparable from voice—and both are inseparable from place. Voices help co-compose atmospheres and, in turn, atmospheres voice themselves. The voices of atmospheres through artworks, in artworks, and around artworks can affectively enliven the experience of the work and the place they are in. And, in a minor gesture, the voices of atmospheres also bring something more-than to the fore—something that speaks the ethics and politics of the place and of art.

What this chapter's particular engagements with art works have proposed, then, is that apprehensions of affective atmospheres have much to offer to voice studies and voice studies have much to say to understandings of atmospheres. Engaging with specific art projects has addressed how voice could be thought of or re-conceptualized as more-than-human, atmospheric, and entangled with new materialist concerns. Simultaneously, listening to atmospheres and voices speak and co-compose these artworks has, I would hope, opened lively, new attention to voice, atmospheres and new materialism itself.

Acknowledgments

I would like to pay my respects and acknowledgments to all Traditional Custodians on whose land I live, work, and travel through. I acknowledge that sovereignty has never been ceded. I would like to thank Joel Stern for generously giving his time for an interview and offering much valuable food for thought. Thank you, too, to Maria Miranda for sharing her paper, "From the Ground Up: artist-run initiatives and place" (AAANZ Conference, December, 2017) and her insights on atmospheres, as well as on our work together with the worms.

Some of the material in this chapter comes from my previous publications:

Maria Miranda and Norie Neumark, "Small 'p' politics and minor gestures: *political* artists, politics and aesthetics in contemporary art" in Graham Meikle (ed.) 2018. *The Routledge Companion to Media and Activism*. London and New York.

Norie Neumark, "Eavesdropping, Curated by Joel Stern and James Parker", *A+a Online*. 2018 http://www.artandaustralia.com/online/discursions/eavesdropping-curated-joel-stern-and-james-parker.

Norie Neumark, 2015. "Enchanted Voices: voice in Australian sound art" in *Voice Studies: Critical Approaches to Process, Performance and Experience*, (Konstantinos Thomaidos and Ben Macpherson, eds) London and New York: Routledge.

Norie Neumark. 2017. *Voicetracks: Attuning to Voice in Media and the Arts*. Cambridge MA: MIT Press.

Notes

1 Similarly in German, Stimme is the same word as voice, and Stimmung as atmosphere. This is particularly noteworthy, given German scholars' influential theorizations of atmosphere, as I elaborate later in the chapter (e.g., Gernot Böhme, "The Aesthetics of Atmospheres," 2017).

2 It is interesting that so far some of the most vital new materialist thinking on voice within sound studies has come from musicologists, who are alert to the more-than of the musical voice. Nina Eidsheim, for instance, explores the collective, cultural production of an 'individual' human voice and the social role of the listener in eliciting it (Eidsheim 2015, 2019). Another musicologist whose new feminist materialist approach has enlivened voice is Milla Tiainen. Her analysis of voice works with a view of "a world where human life and cultural practices, technological procedures, and other-than-human natures are always-already interlocked in technoecological relations" (Tiainen 2017).

3 I expand on this in Neumark 2017, Chapter 1. I would note, too, J. Martin Daughtry's insightful understanding of *auditor* not just as listener but also as inflected by the call to *audit*, to take responsibility. *Auditor* in this double sense speaks of the need to call to account—to make an *audit* of sonic forces, in the case of his research, sonic violence (Daughtry 2015).

4 I should note there is quite a different, more sympathetic and very important approach to ambiance by Jean-Paul Thibaud (and others who work with his idea of ambiance)—as a situated, atmospheric sensitivity to the world—an approach along the same lines as Böhme, with whom Thibaud collaborates. (For instance, see Atmospheric Spaces: Aura Stimmung Ambiance and Thibaud 2011a, b.)

5 Jane Bennett, too, is drawn to Thoreau's work, which she reads through her ethical concerns and her attention to enchantment. She is attuned to Thoreau's sensibility, including "a deliberate and vivid way and practices of seeing, hearing, touching, smelling, and moving," which he cultivates through "a series of techniques of the self" (Bennett 2001: 95. See also 92–94 and Bennett 2010: 45–49).

6 Cultural critic Sian Ngai has developed a number of other, potentially useful analyses of minor feelings, performances and processes, such as 'interesting' and 'gimmick'. However, since I perceive in Stern's work neither *interesting*'s notability but lack of intensity nor *gimmick*'s object form, I prefer just to work with zany, the figure through which Ngai particularly reads performances.

7 This was long ago figured by François Rabelais, whose Panurge and Pantagruel encountered frozen words and deeds—an encounter famous within sound studies and prescient for new materialism. This is an enchanted encounter—where voice and sound are visible and palpable to the touch and where words frozen in winter melt in spring into noisy voice and action. It is so full of potential that "at any moment" it may be actualized—or perhaps more than, different from actualized—listened to in its present/absent fullness.

8 Previously, we wrote about this as 'collaboration' but now, in recognition of the fact that we never sought nor received the worms' consent, we prefer to say 'work with' and 'follow,' following Jacques Derrida and Katve-Kaisa Kontturi. Tellingly, the very title and text of Derrida's book plays with the French for *je suis,* which means **both** "I am" and "I follow" (Derrida 2008).

9 The video is available at https://vimeo.com/247735081
I discuss sonification in *Voicetracks* (2017: 96–97), referring to sound theorist Jonathan Sterne's cogent critique. I also respond to cultural historian Stephen Connor's recognition of the human-centered urge to render sound as 'for us' and 'in our terms.' *Our terms* ... yet again an assertion of the human as central. As Connor alerts us, sonification does not invite or require us to bend to hear the

voice of others, things, events, but instead bends them to speak in a voice we prefer to hear, a voice we are accustomed to listening to. It is as if we need to sonify in order to hear the voices of objects, rather than attuning ourselves differently.

References

Art projects referenced

Hamdan, Lawrence Abu. 2018. *Rubber Coated Steel.* Melbourne: Ian Potter Museum of Art.
Miranda, Maria, and Norie Neumark. 2018. *Worms & Worms & Worms.* Trocadero Artspace, Melbourne. 29 August – 15 September. http://www.trocaderoartspace.com.au/maria-miranda/august-29-september-15-gallery-two-maria-miranda-worms-worms-worms/
Neumark, Norie, and Maria Miranda. 2017. *Waiting.* Melbourne: MoreArt.
Stern, Joel. 2017. *I Am Sitting in a Courtroom, in Acoustic Justice.* Melbourne. https://youtu.be/XoSrykqa4So

Texts referenced

Anderson, Ben. 2009. "Affective Atmospheres." *Emotion, Space and Society* 2: 77–81.
Anderson, Ben, and James Ash. 2015. "Atmospheric Methods." In Phillip Vannini (ed.). *Non-Representational Methodologies: Re-Envisioning Research.* New York and London: Routledge.
Atmospheric Spaces: Aura Stimmung Ambiance. https://atmosphericspaces.wordpress.com/
Barad, Karen. 2012. "On Touching—The Inhuman that Therefore I Am." *Differences* 23 (3): 206–223.
Barad, Karen. 2015. "Ecologies of Nothingness and Troubling Time/s: On Devastation and the Im/possibilities of Endurance." Transversal Practices: Matter, Ecology and Relationality. VI International Conference on New Materialisms. Victorian College of the Arts, University of Melbourne.
Barrett, Estelle, and Bolt Barbara, eds. 2013. *Carnal Knowledge: Towards a 'New Materialism' through the Arts.* London: I.B. Tauris.
Bennett, Jane. 2001. *The Enchantment of Modern Life: Attachments, Crossings and Ethics.* Princeton: Princeton University Press.
Bennett, Jane. 2010. *Vibrant Matter: A Political Ecology of Things.* Durham: Duke University Press.
Böhme, Gernot. 1993. "Atmosphere as the Fundamental Concept of a New Aesthetics." *Thesis Eleven* 36: 113–126.
Böhme, Gernot. 2013. "The Art of the Stage Set as a Paradigm for an Aesthetics of Atmospheres." *Ambiances.* Redécouvertes, mis en ligne le 10 février 2013. http://journals.openedition.org/ambiances/315; https://doi.org/10.4000/ambiances.315 (consulté le 15 juillet 15, 2025).
Böhme, Gernot. 2017. "Chapter 2. The Ecstasies of Things: Ontology and Aesthetics of Thingness." In A.-Chr. Engels-Schwarzpaul (ed. and trans.). *Atmospheric Architectures: The Aesthetics of Felt Spaces.* London and New York: Bloomsbury Academic.
Braidotti, Rosi, and Posman Sarah. 2013. "Conversation with Rosi Braidottti." *nY* 19/06/2013. https://www.ny-web.be/artikels/conversation-rosi-braidotti/ (Accessed July 15, 2025).
Connolly, William. 2013. "The 'New Materialism' and the Fragility of Things." *Millennium: Journal of International Studies* 41 (3): 399–412.

Darwin, Charles. 2007. *The Formation of Vegetable Mould through the Action of Worms, with Observation of Their Habits*. Teddington UK: Echo Library.
Daughtry, J. Martin. 2015. *Listening to War: Sound, Music, Trauma, and Survival in Wartime Iraq*. Oxford: Oxford University Press.
Derrida, Jacques. 2008. *The Animal that Therefore I Am*. trans. D. Wills. New York: Fordham University Press.
Despret, Vinciane. 2004. "The Body We Care For: Figures of Anthropo-zoo-genesis." *Body and Society* 10 (2–3): 125–170.
Despret, Vinciane. 2013. "Responding Bodies and Partial Affinities in Human-Animal Worlds." *Theory, Culture and Society* 30 (7/8): 51–76.
Dolphijn, Rick and Iris Van der Tuin. 2012. *New Materialism: Interviews and Cartographies*, Open Humanities Press. ISBN: 9781607852810.
Eidsheim, Nina Sun. 2015. *Sensing Sound: Singing and Listening as Vibrational Practice*. Durham: Duke University Press.
Eidsheim, Nina Sun. 2019. *The Race of Sound: Listening, Timbre, and Vocality in African American Music*. Durham: Duke University Press.
Ihde, Don. 2007. *Listening and Voice: Phenomenologies of Sound*. 2nd ed. Albany: State University of New York Press.
Ingold, Tim. 2011. *Being Alive: Essays on Movement, Knowledge, and Description*. London: Routledge.
Haraway, Donna. 2008. *When Species Meet*. Minneapolis: University of Minnesota Press.
Harward-Nalder, Glenda. 2018. "Why Listen to Plants? (A Plant Is a Community) – An Australian First Nations Perspective." Liquid Architecture https://liquidarchitecture.org.au/events/why-listen-to-plants
Kahn, Douglas. 2013. *Earth Sound Earth Signal: Energies and Earth Magnitude in the Arts*. Berkeley: University of California Press.
Kim-Cohen, Seth. 2016. *Against Ambience and Other Essays*. London and New York: Bloomsbury.
Kontturi, Kaisa-Katve. 2018. *Ways of Following: Art, Materiality, Collaboration*. London: Open Humanities Press.
Latour, Bruno. 2004. "How to Talk about the Body? The Normative Dimension of Science Studies." *Body and Society* 10 (2–3): 205–229.
Liquid Architecture and Melbourne Law School, 2017. *A courtroom is not a gallery, Acoustic Justice Project*. https://liquidarchitecture.org.au/events/acoustic-justice
Manning, Erin. 2016. *Minor Gestures*. Durham, NC: Duke University Press.
Manning, Erin. 2010. "Always More than One: The Collectivity of *a* Life." *Body & Society* 16 (1): 117–127.
Marder, Michael. 2013. *Plant-Thinking: A Philosophy of Vegetal Life*. New York: Columbia University Press.
Neumark, Norie. 2010. "Introduction: The Paradox of Voice." In Norie Neumark, Ross Gibson, and Theo van Leeuwen (eds.). *Voice: Vocal Aesthetics in Digital Arts and Media*. Cambridge, MA: MIT Press.
Neumark, Norie. 2017. *Voicetracks: Attuning to Voice in Media and the Arts*. Cambridge, MA: MIT Press.
Ngai, Sianne. 2015. *Our Aesthetic Categories: Zany, Cute, Interesting*. Cambridge, MA: Harvard University Press.
Rancière, J. (2004) *The Politics of Aesthetics: The Distribution of the Sensible*. Trans. and introd. Gabriel Rockhill. London and New York: Continuum.
Rockhill, G. (2004) "Translator's Introduction: Jacques Rancière's Politics of Perception" in Rancière, J. (2004).
Stern, Joel. Interview with the author. 2018.
Stewart, Kathleen. 2011. "Atmospheric Attunements." *Environment and Planning D: Society and Space* 29: 445–453.

Stewart, Kathleen. 2012. "Precarity's Forms." *Cultural Anthropology* 27 (3): 518–525.
Thibaud, Jean-Paul. 2011a. "The Sensory Fabric of Urban Ambiances." *Senses and Society* 6 (2): 203–215.
Thibaud, Jean-Paul. 2011b. "A Sonic Paradigm of Urban Ambiances." *Journal of Sonic Studies* 1 (1). https://www.researchcatalogue.net/view/220589/220590
Thomaidis, Konstantin. 2018. "Voice: Performance and Forensics." In Experience Bryon (ed.). *Performing Interdisciplinarity: Working across Disciplinary Boundaries through an Active Aesthetic.* Oxon and New York: Routledge.
Tiainen, Milla. 2017. "Sonic Technoecology: Voice and Non-Anthropocentric Survival in *The Algae Opera*." *Australian Feminist Studies* 32 (94). Published online May 2018. https://doi.org/10.1080/08164649.2017.1466651
Voegelin, Salomé. 2010. *Listening to Noise and Silence: Towards a Philosophy of Sound Art.* New York: Continuum.
Weizman, Eyal. 2014. "Introduction," *Forensis: The Architecture of Public Truth* (Ed. Forensic Architecture). Berlin: Sternberg Press. http://www.forensic-architecture.org/wp-content/uploads/2014/04/Forensis-interior-FINAL_Weizman-Intro-I.pdf (Accessed July 24, 2018).

Afterword

Laura Cull Ó Maoilearca

Consider a book as echolocation.
Consider a book as a process of sounding and listening to the echoes.
Consider a book as a practice of call and return.

Ontological echoes: Voicing new spiritualism

What response might we send back to the calls of *Mattering Voices*? This book invites us to consider how the intra-actions between transdisciplinary voice studies and new materialisms might "enrich understandings of how voices matter – how they materialize into diverse events and practices and come to mean different things" (Tiainen, Belgrano and Tarvainen in this volume). Among its many gifts, it is a welcome feature of this book that the emphasis lies on the "materialisms" more so than on the "new" – given that ontologies and practices attending to the vital strangeness of matter not only have centuries of European heritage but were, in many respects, already there in First Nations, Indigenous philosophies. For instance – as a first echo – my own thinking on voicing emerges more from the context of process philosophical approaches to materiality, rather than specifically new materialist ones – pointing toward the precursors of this thinking as far back as at least the 19th-century thought of Henri Bergson and others. Here, "the physical world is understood as movement, energy, and force" (Ó Maoilearca 2023: 8). Matter is construed as tendency rather than substance; and, akin to Barad, the emphasis is on the notion of matter as a process of materialization. In Bergson's *Matter and Memory,* for instance, perception is considered as different from matter only in degree rather than kind, according to a part/whole relationship. And indeed, there is no dualism of matter and spirit in this thinking or in the wider tradition of "French Spiritualism" in which it might be situated; rather, the place of spirit is retained "through *movement*, *duration*, and *habit*" (Ó Maoilearca 2023: 14).

In this respect, a further welcome quality of this book's call is that it does not eschew spirit – indeed, it begins with it: with "the materialisation of spirit and the spiritualisation of matter" (Belgrano and Price in this volume), and with the counter-intuitive thematic triad of voice, new materialism and nothingness. Starting as it means to go on, this opening act makes clear that this volume is not just a matter of new materialism informing voice studies, but also – critically – the other way around. After all, as John Ó Maoilearca writes in *Vestiges of Philosophy*, "the possibility of a *nontranscendent* (or immanent) spirit is rarely, if ever, entertained" in new materialism (Ó Maoilearca 2023: 13). Or as Belgrano and Price note in their chapter: many new materialists "are opposed to any talk of 'spirits' whatsoever" (Belgrano and Price). Through its antipathy to spirit, there is still an exclusionary tendency within much new materialism, despite the extension of life to matter and the expansion of agency to the more-than-human. So, in continuing this work beyond these pages – or in sending back soundings in return – perhaps we can ask: how might [new] materialist approaches to voice continue to explore their relation to concepts and practices with historically immaterial connotations: the spiritual, psychical, esoteric, magic, gnosticism, hermeticism, occultism, and mysticism? How might we expand the consideration of material voicings to ask how the past speaks, how the dead speak as mattering voices that differ in degree rather than kind from others? How do we participate in voicings that do not issue from some "other" transcendent realm or parallel reality but from the other temporal realities nested within this world? What extra-ordinary practices of perception and listening might be trained to hear these mattering voices?

There is a beautiful indication of this expansive direction – taking new materialism beyond itself – already in Chapter 1 of this book, where I hear echoes of Bergson in Belgrano and Price's attention to the material presence of the past. As they write:

> Voice can call forth the presence of the past – not as mere nostalgia but as *live* performance and real material connection with spatial and temporal 'otherness'. Song bursts out of the logic of linear time and self-identity. It can disrupt the usual cognitive systems which operate in terms of either presence *or* absence.
>
> (Belgrano and Price, in this volume)

As a divergent echo, I think again of *Vestiges*, where Ó Maoilearca discusses the esoteric practices of Bergson's sister – the artist and mystic Mina Bergson. Here, he describes her use of voice in the context of the rituals of the Golden Dawn, including through a vibratory method of chanting, which emphasizes the material performativity of sound. As he describes:

> After all, "chanting," the rhythmic repetition of voice, shares its etymology with "enchantment" (from *cantare*, to sing) wherein the world is brought under the influence of magic. Instructions for the "correct pronunciation and vibration" of divine names (often from the Enochian "angelic" language)—for *water*, say—appear repeatedly in Mina Bergson's texts. Pronunciation is a careful *material sounding* that must

> be performed correctly if it is to be effective ... Mina Bergson and the Golden Dawn used vocal repetition (chanting) to enchant the voice (making it a divine communication).
>
> (Ó Maoilearca 2023: 107)

Consider a book as echolocation.
Consider a book as a process of sounding and listening to the echoes.
Consider a book as a practice of call and return.

Interspecies echoes: Voicing intersectional solidarities

A second echo. As much as aesthetics and onto-epistemology, this book also indicates how questions of voice are fundamental to ethico-political concerns – to questions of which materialities are understood to have a voice, which voices matter, and practices of "giving and taking voice" within processes of democratization, oppression and liberation (McCaffrey 2023). Some of the recent advances in animal politics, for example are connected to the production of discourse that challenges historic assumptions about the relationships between non-human animals and language (and notions of "voice") and the impact of this on notions of political representation. For example, Dutch animal philosopher Eva Meijer's work investigates "how existing political practices and institutions can be extended to incorporate non-human animal political voices, and to explore new ways of interacting with other animals politically" (Meijer 2019: 12). Even just the title of her book – *When Animals Speak* – tells us that this thinking moves away from a political model where humans are tasked with giving voice to non-humans animals understood as beings who cannot speak for themselves, toward a model based on the idea that animals do indeed already speak in their own ways – and where the new task for humans is to acknowledge animals as indeed speaking and having voices that have political interests and consequences, and to get better at listening to what they are saying. Moving away from the use of language as a marker of human exceptionalism and measuring animal language ability according to human standards, there is a growing acceptance of the notion of animal languages and that animals speak – both in animal science and indeed in (socio)linguistics (such as in the important work of Leonie Cornips), which has historically been exclusively concerned with human languages.[1] And yet, at the same time, Meijer notes that:

> Many animal rights theorists [still] simply assume that other animals do not use language, cannot speak in a way that is relevant to political action, and are destined to stay silent in political and social matters. There are very few animal rights theorists who incorporate the current research on non-human animal, or interspecies, languages in their work ... Activists and philosophers currently often see it as unavoidable that "we" should speak for "them". This is unfortunate ... because it

> perpetuates a negative stereotype of other animals as mute, or incapable of speaking, and keeps intact an idealized view of the human and human language.
>
> (Meijer 2019: 4)

So, what can a new materialist approach offer to this conversation? Clearly, it has much to bring to the question of interspecies voice or how voicings are constituted through interspecies encounters, as discussed so beautifully in Jennie Tiderman-Österberg's chapter on herding musicking in this volume. Thinking alongside the "with-ness" of the cow choirs, I immediately think of the vocal improvisation with farm animals that happened in the context of the *Sheep Pig Goat* project by the UK arts company Fevered Sleep which I collaborated on in 2017–2020. *Sheep Pig Goat* is a "creative research studio" that explores the questions of what animals have to say "through a series of improvised encounters between human performers and sheep, pigs, goats and other livestock animals" (Fevered Sleep 2024). Here, performers – including vocal artists – were invited to pay careful attention to the vocalizations of non-human animals as expressive, relational and perceiving beings, rather than as mere objects of anthropocentric knowledge. How might voicing practices enable us to unlearn the anthropocentrism that compels us to hear the sound of *a* sheep rather than the distinctive vocalization of *this* ewe (in all her vocal singularity in relation to her lamb)? For instance, the vocalist Sterre Maier used her expertise to attend to the animals' attention to voice and breath: to notice the pitch and quality of sounds that appeared to soothe the sheep and those that "hold them a little bit more in focus", those that seemed to cause them to step away and those that brought them toward her (Maier in Cull Ó Maoilearca 2024). She was not singing *to* the sheep according to a voice that she already possessed; the relationality between Maier, the sheep, the space and other materialities produced an interspecies event of voicing that attunes us to both the creative and oppressive dimensions of dispossession.

Throughout the texts in this volume, authors explore voice in terms of new materialist approaches with a focus on two key ideas. In the first instance, there is the emphasis on a relational ontology which foregrounds voice understood as processes of voicing constituted by relations, rather than by the notion of voices as issuing from and expressing the identities and intentions of subjects. Second, the texts emphasize the constitutive or agential power of more-than-human entities and materialities in this relational production of voicings.[2] In this context, one exciting potential of the [new] materialist approach is the specific invitation it offers to take a more-than-human approach to notions of intersectionality within a social justice framework. On the one hand, the [new] materialist approach makes clear that attending to the intersecting dynamics of oppression and liberation need not be a matter of "identity politics", but a question of the emergent production of subjectivities from overlapping, constitutive processes of gendering and

Figure A.1 Fevered Sleep Sheep Pig Goat 2017 Sterre and goat.
Image credit: Ben Gilbert for Wellcome Collection.

racialization – for example. It can draw our attention to the dynamics of micro and macro: how structural oppressions and creativity operate in the context of unique assemblages of materialities in specific evental contexts. On the other hand, the more-than-human dimension might also allow us to consider how (de)humanization might also be part of this process – opening to questions of, for instance, where social justice and animal justice meet. The politics of animal voicing is not separate from the politics of human voicing insofar as a normative standard of who counts as (truly) human – and therefore attributed to a voice – operates across interspecies relations. As we continue with this work, perhaps we can ask: Where do practices of voice and voicing sit within these intersectional dynamics of a more-than-human justice? How do questions of the enrichment or undermining of trans* senses of self among humans (as per Tiainen's chapter in this volume) – for example – intersect with the manifestation of animal agency? How might [new] materialist approaches help us to understand the shared logic of normative notions of which voices count or which voices count as "human" which manifests across transgender, racialized and interspecies voicings?

In terms of transness, Eva Hayward's "More Lessons from a Starfish" (2008) already shows us a way here, where the song "The Cripple and the Starfish" by Anohni and the Johnsons is an agential force within the relational emergence of transsexual, transgender and transspecies selfhood. As Hayward puts it: "in the text of Anohni's voice, in the instrumental variations, and in the lyrics themselves, boundaries of sexual and species differences, artificial and authentic orderings, and nature and culture are affectively and literally *trans*-ed" (Hayward 2008: 69). Alongside this work, I also think of the "wayward" moaning practice of the Amsterdam-based artist Raoni Muhzo Saleh who writes of "lending my voice, my shrieks, my moans, as a way of amplifying the voices of those *violently silenced*": not comparing, but voicing the interdependence of whales wailing, "trans-moans" and the cries of grief and rage that emerge from colonial violence (Saleh 2024).[3] As Saleh asks through his practice and informed by Saidiya Hartman: "How do we learn to become intimate with one another by way of sounding together, through the difference of our dispossessions without separating?" (Saleh 2023).

In terms of the intersections of speciesism and racism, I think of the *Whale Whispering* (2018–ongoing) project of Michaela Harrison which so powerfully enacts and invites reflection on how voices matter and how voices act in the entangled histories of coloniality, the transatlantic slave trade and the more-than-human life of the ocean. A US-based musician of African descent, Harrison performs internationally as a singer who describes her work as "rooted in relaying the healing, transformational power of music through song". In 2018, *Whale Whispering*, Harrison began "going out to sea to sing with the [humpback] whales and co-create songs with them" – starting with singing spirituals to them, but moving toward "an emphasis on wordless singing and spontaneous composition, co-creation" (Harrison in Bazil 2020).

Returning to the question of what happens to "spirit" or "the spiritual" in new materialism, it is worth noting that Harrison describes her motivation for connecting with the whales as relating to: "evolving my relationship with my ancestors and understanding how I have been carrying forth legacies, of trauma, resilience and victory mentality, through my connection with them" (Harrison in Bazil 2020). For Harrison, *Whale Whispering* "is related to the Middle Passage (when enslaved Africans were transported to the Americas as a part of a triangular trading route between Europe, Africa and the Americas) ... How did anybody survive that even for a day? Singing is one of the things that we've done throughout this trajectory that has gotten us through and gotten us over" (Harrison in Bazil 2020). Singing is survival. In turn, talking about her practice, Harrison then speculates about a song exchange between whales and enslaved Africans during the Middle Passage, suggesting that enslaved Africans "heard the whales singing and the whales heard them singing, and there was this exchange that happened" (Harrison in Bazil 2020). She also notes how interspecies collaboration may have taken place in this context, since "whales ... incorporate songs from the environment, as well as the sounds from the environment, into their songs" (Harrison in Bazil 2020). *Whale Whispering* returns us to matter-spirit tendencies then, but also to the idea of nested temporalities: to voicings emerging "in the middle of many rhythms" (Tiainen in this volume).

Going further with this assemblage of voicing: Harrison's work engages in interspecies and ancestral connection not only through song but also through site. As a project, *Whale Whispering* is conducted in collaboration with Institute Projeto Baleia Jubarte, based in Praia do Forte, Bahia, Brazil. If you google it, you'll see that Praia do Forte is a beach and small coastal fishing village, now "being developed as an ecologically minded upscale beach resort". But what you won't find so easily are the entangled histories of the site that drew Michaela Harrison to this place and informed her vocal practice: that Praia do Forte was both the site of whale hunts and a massive plantation during colonization and enslavement. Performing on-site, Harrison's concerts emerge from the entangled histories of black slavery and animal exploitation. But also – as she notes – the dynamics of renewal since now, Bahia is also "the place where the whales come to give birth. They come to Bahia to reproduce and then they migrate to Antarctica to feed" (Harrison in Bazil 2020).

Consider a book as echolocation.
Consider a book as a process of sounding and listening to the echoes.
Consider a book as a practice of call and return.

(Taking the Castle)
(by Michaela Harrison)

Figure A.2 Michaela Harrison.
Image credit: Ismael Silva.

What else is there to do? All these bones have been coming to me, coming to me, asking
me to let their stories be sung through me and I hear the hum of their dna in the ruins of
the castle calling for their memory to be rung
from the top of the hill where all this water can be seen –
a bell, a bell, a siren's peal
a wail, a whail,[4] a whalesong
for the egun,
human and baleen.
Because it's all happening now, my offerings for the ancestors fortify my current, embodied
self via their blood rivering through my veins
its pulse echoing their refrains
I don't recall their names yet
I can never forget their voices.
Those who sailed here in the wake of so much treachery
who lunged and lurched out on the open omí[5]
cried from the outset for the Mother of the Fishes knowing she would cradle their orí[6]
in Love, so she grew
to cover the vastness between two lands that had once stood hand-in-hand,
answered their prayers
with lullabies from outside the hull
those melodies moving with the whales
their escorts to the ports of hell
singing them proof that at every turn, exquisiteness can still be found.

(Harrison 2024)

Notes

1 Reading Tiderman-Österberg's text, I also hear the echoes (or rather, the mooing) of the work of the Dutch cow language expert, Leonie Cornips, and her recent dialogical collaboration with the artists Mark Jeffery and Lucy Cash; and scholar Louis van den Hengel. For more on Cornips work, see this chapter of the forthcoming edited collection *Interspecies Performance* (2024) co-edited by myself and Florence Fitzgerald-Allsopp.

2 In my own work, I tend to focus on non-human animals – rather than on the notion of the more-than-human – and with a specific interest in the notion of the non-human understood not as negation but as a qualitative expansion of the human (in parallel to Laruellian non-philosophy as outlined by John Ó Maoilearca, Katerina Kolozova and others, and also referred to by Nauha in this volume).

3 See also sonic performances by Saleh in collaboration with composer Abel Kroon such as *Trans(in) moan(in)* (2023). As Saleh describes: "This performance deals with the question: how do we learn to become intimate with one another by way

of sounding together, through the difference of our dispossessions without separating?" (Saleh 2023, np).

4 Harrison states: "Whail is a word I made up combining whale and wail to speak to describe the practice of singing and screaming/bellowing/wailing with whalesong as a form of healing release of personal and ancestral trauma" (Harrison 2024).

5 Harrison states: "Omi is water in the Yoruba language" (Harrison 2024).

6 Harrison states: "Ori, used here in the plural sense, is the Yoruba word for the physical head, the mind, and also refers to the spiritual consciousness/divine self of a person" (Harrison 2024).

References

Bazil, Madeleine. 2020. Whale Whispering. Interview by Madeleine Bazil with Michaela A. Harrison, available at https://wheretheleavesfall.com/explore/article-index/whale-whispering/ (accessed 07/03/24).

Cull Ó Maoilearca, Laura. 2024. Performing Knowledge as Interspecies Encounter: Sheep Pig Goat and the Ethics of Attention. In: Cull Ó Maoilearca, Laura and Florence Fitzgerald-Allsopp (Eds.). *Interspecies Performance*. Aberystwyth, Wales: Performance Research Books.

Fevered Sleep. 2024. *Sheep Pig Goat*, company website https://www.feveredsleep.co.uk/project/sheep-pig-goat (accessed 07/03/24).

Harrison, Michaela. 2024, forthcoming. Whale Whispering. In: Cull Ó Maoilearca, Laura and Florence Fitzgerald-Allsopp (Eds.). *Interspecies Performance*. Aberystwyth, Wales: Performance Research Books.

Hayward, Eva. 2008. More Lessons from a Starfish: Prefixial Flesh and Transspeciated Selves. *Women's Studies Quarterly*. 36 (3/4), 64–85.

Meijer, Eva. 2019. *When Animals Speak: Towards an Interspecies Democracy*. New York: New York University Press.

McCaffrey, Tony. 2023. *Giving and Taking Voice in Learning Disabled Theatre*. London/New York: Routledge.

Maoilearca, John, Ó. 2023. *Vestiges of a Philosophy: Matter, the Meta-Spiritual, and the Forgotten Bergson*. Oxford: Oxford University Press.

Saleh, Raoni Muzho. 2023. 'Trans(in) Moan(in)', Performance Description, available at https://www.uva.nl/en/shared-content/faculteiten/en/faculteit-der-geesteswetenschappen/events/2023/06/transin-moanin.html (accessed 07/03/24).

Saleh, Raoni Muzho. 2024, forthcoming. Praxis Moaning while in Overwhelm. *With the Dead: Performance Philosophy, Grief and Dying*, Special issue of the *Performance Philosophy* journal, Volume 9, Number 2 (November 2024), edited by Laura Cull Ó Maoilearca, Will Daddario and Rajni Shah.

Index

Note: Page numbers in italics indicate figures, and references following "n" refer notes.

FSC
www.fsc.org
MIX
Papier | Fördert
gute Waldnutzung
FSC® C083411

Kolibri 360 GmbH
Ferdinand-Jühlke-Straße 7
99095 Erfurt, Deutschland
produktsicherheit@kolibri360.de